Routledge Revivals

The Spirit and Purpose of Geography

First published in 1951, *The Spirit and Purpose of Geography* offers an introduction to the scope and spirit of geography. This undertakes a no less ambitious task than that of discovering the spatial relationships of the manifold features, physical and human, which diversify the Earth's surface.

The authors one of whom first approached the subject from physical science, and the other from social science, co-operate to define and to discuss the historical development of their subject, its fundamental physical basis, its cartographic methods, its human aspect and its many applications and problems. Above all they submit that geography, the study of country or landscape, as a link study between the natural sciences and the humanities, constitutes not only a worthy academic discipline but also a part of a liberal education. This introductory volume is a must read for any student of geography.

The Spirit and Purpose of Geography

S. W. Wooldridge and W. G. East

First published in 1951
by Hutchinson's University Library

This edition first published in 2024 by Routledge
4 Park Square, Milton Park, Abingdon, Oxon, OX14 4RN

and by Routledge
605 Third Avenue, New York, NY 10017

Routledge is an imprint of the Taylor & Francis Group, an informa business

Publisher's Note

The publisher has gone to great lengths to ensure the quality of this reprint but points out that some imperfections in the original copies may be apparent.

Disclaimer

The publisher has made every effort to trace copyright holders and welcomes correspondence from those they have been unable to contact.

A Library of Congress record exists under LCCN: 51008732

ISBN: 978-1-032-86176-0 (hbk)
ISBN: 978-1-003-52173-0 (ebk)
ISBN: 978-1-032-86179-1 (pbk)

Book DOI 10.4324/9781003521730

THE SPIRIT AND PURPOSE OF GEOGRAPHY

by

S. W. WOOLDRIDGE, D.SC.
PROFESSOR OF GEOGRAPHY,
UNIVERSITY OF LONDON,
KING'S COLLEGE

AND

W. GORDON EAST, M.A.
PROFESSOR OF GEOGRAPHY,
UNIVERSITY OF LONDON,
BIRKBECK COLLEGE

LONDON
HUTCHINSON'S UNIVERSITY LIBRARY

Hutchinson & Co. (Publishers) Ltd.
London Melbourne Sydney Auckland
Bombay Johannesburg New York Toronto

First published 1951
Reprinted 1952
Reprinted 1955
Reprinted 1956

Printed in Great Britain
at the Gainsborough Press, St. Albans,
by Fisher, Knight and Co. Ltd.

Amongst others, I was then encountered, on my passage from Westminster to Whitehall, by a tall big gentleman, who thrusting me rudely from the wall, and looking over his shoulder on me in a scornful manner, said in a hoarse voice these words: *Geography is better than Divinity*; and so passed along.

DR. PETER HEYLYN,
Cosmographia, preface (1649).

CONTENTS

LIST OF SKETCH-MAPS

PREFACE

"I WOULD have her instructed in Geography," said Mrs. Malaprop, "that she may know something of the contagious countries." This precept has now an added force—in a world which is fast becoming one neighbourhood. But what kind of geography? For, as R. H. Tawney once put it: "there are as many ways of writing geography as of writing history."

The study and writing of geography began long ago: veins of geography thread even the fabulous *Odyssey* of Homer. But as a coherent and autonomous discipline worthy to rank with old established university studies, it is relatively a new-comer, especially in this country. It came late as a modern science because its re-birth awaited the remarkable advances in the nineteenth century of the natural and social sciences, which alone could provide that background knowledge of the earth, without which geography would have remained very much a matter of wondrous speculation. It has now come of age in this country, and its shape and purpose grow increasingly clear. But it is still at an early stage of its development and its practitioners are following many paths which might appear, at first sight, divergent. For such reasons the time has seemed opportune to try and see geography whole, to lay bare its foundations, to indicate its special aspects, to make clear its objectives and at least to suggest its many applications.

Our small book is offered as an introduction to the scope and spirit of geography. This undertakes a no less ambitious task than that of discovering the spatial relationships of the manifold features, physical and human, which diversify the earth's surface. We have tried here to signpost a path which can lead the student, if he is so minded, towards the full study of his heritage, the earth's surface which he treads. And if he perseveres along this path, his journey will not prove unrewarding. For his effort will subject him to a discipline and yield him a philosophy.

S. W. W.
W. G. E.

London,
July, 1950

CHAPTER I

THE NATURE AND DEVELOPMENT OF GEOGRAPHY

It is one of the glories of modern knowledge that the study of geography has been transformed. . . . The proper study of man's habitat is one of the triumphs of the rational spirit.

K. B. SMELLIE, *Why We Read History* (1947).

The Concise Oxford Dictionary declares soberly under "geography": "Science of the earth's surface, form, physical features, and political divisions, climate, productions, population, etc."

This definition is true without being either very helpful or illuminating to the serious student of the subject. Even this brief statement would make it appear that the scope of geography is distractingly wide and its aim far from clear. The form of this earth is strictly the concern of geodesy, its physical features are, in part at least, the concern of the geologist, and its climates the result of meteorological processes, the study of which is a branch of applied physics. No less does the explanation of its political divisions appear to fall to the historian, since they are the outcome of long-term human processes—migrations, wars, revolutions and the complex sequence of political and social change. "Production," it is true, is in one aspect an affair of geography, but it is also more especially the concern of economics. Similarly, although the distribution of population is plainly of geographical interest, much has been written on population and its problems which is outside both the interest and the competence of the geographer.

The only phrases in the dictionary definition on which we have made no comment are those which pre-empt for geography the "earth's surface" and its "natural divisions." It might well be possible to limit the definition to these, with suitable elaboration and reservation, but all the features, factors or phenomena which seem to pertain to other studies would, in fact, still be involved. Nor is the above list by any means exhaustive; in relation to the earth's plant cover and its agriculture, botany at least must be admitted to the group of subjects sharing the field, and the other divisions of biological science are only less important.

Herein lies the peculiar virtue, or as some hold, the inherent vice of geography regarded as a "subject." It fuses the results, if not the methods, of a host of other "subjects" and in its full latter-day development seems to require a knowledge of a larger range of ancillary studies than almost any other science or art. The man, it might seem, who, with his brief span of three score years and ten, sought to qualify as a geographer would perish, like Browning's Grammarian, long before he reached the end of his interminable academic trail and certainly before he had reached geography. Hence the view of the layman that geography is not a science but merely an aggregate of sciences, and the extreme pessimism of a Polish geographer who was afraid to devote all his life to the study of geography lest he should find at the end that, as a science, it did not exist!

There is little doubt that this difficulty or some aspect of it has been experienced by every serious student of the subject. In the Universities, geography is still young. For many years, in Britain at least, the learned world refused to admit that such a subject could exist in any real or valuable sense or, at least, that it could be pursued beyond the elementary school stage. In its higher levels it appeared to fall between the fields of geology and history and to be permissible, if at all, only under their suzerainty. It is of course true, that an individual can acquire by travel and study a large compendium of knowledge about the world—"the earth's surface"—but it is implicitly doubted whether such knowledge can become "organized knowledge" worthy to rank as a discipline beside those of older establishment.

Although the aim of this book is to help to lay such doubts in the minds of the student and the general reader, we shall not proceed by way of an immediate frontal attack on the critics of the subject. The matters at issue must be further examined and they cannot be fully treated without "getting down to cases." It may suffice for the present to note the conclusion we shall ultimately reach, that the ordinary man's attitude to the subject, what has been called his "unspecialized curiosity" about the world he inhabits, is in some respects essentially saner than that of those narrow specialists who have denied and still deny the claims of geography.

It has been possible to trace in considerable detail the early history of geographical thinking. None the less a great gulf is set between modern geography and that of classical and mediæval times. It is useful, however, to recall that man, as an observing, comparing and reflecting animal, has always been in some sense a geographer. In the remote past his world or environment was spatially a narrow one; it remains so still for only too many. But it was necessary to cultivate a certain practical awareness of environment—a horse-sense for country—in order to live and survive on any part of our planet. Maps were mental maps, crude and inaccurate, the practical "know how" of the "way about," until comparatively recent times. Yet neither the individual nor the human community could dispense with such maps. The phenomena placed in review were relatively few and perhaps little co-ordinated; there a spring or ford, there a field or forest and, most important of all, the shape of the surface, the complex and ever-varying network of hills and valleys. All these formed a definite pattern, committed to memory and used in living, a rule-of-thumb geography.

Nor need we, with our cartographic technique and scientific learning, despise the "eye for country," the natural geographic sense, of the savage. Many an explorer has been glad to avail himself of it, if indeed he has not often owed to it his life. Such faculty is lamentably lost by the modern tube-travelling suburban dweller; few Londoners are aware, save in the most limited sense, of the pattern of their great city, still less of the qualities and characteristics of the ground upon which it stands and of the texture and meaning of its framework of countryside. But the countryman, to this day, builds up a peculiarly full and satisfying knowledge of his physical environment, of hill and valley, stream and soil and even of the local vagaries of climate.

It may well seem, however, that any attempt to relate primitive "eye for country" to the formal and sometimes sterile abstractions of the geography text-book is playing with words and avoiding realities. Naturally, in so far as geography was and is a practical art, we can descry its origins and perceive its primitive uses just as we might those of, say, metallurgy or economics. But little light on the complexities of either modern metallurgy or economy is cast by its presumed prehistoric origins and it may seem that we gain little more by enunciating the truism that man has always been a geographer. Yet the point is worth making and is essential to the case we are trying here to present.

What we have to face and to reckon with is the enormous growth in the scope of the subject, which began with the Great Age of Discovery and was effectively completed by the end of the 19th century. This growth was twofold, comprising the onward march of exploration and cartography and the birth and development of the specialist sciences. Let us glance briefly at each of these great expansions.

Before the 4th century B.C. the world known to Europeans consisted essentially of the lands surrounding the Mediterranean Sea, together with South-west Asia. Only the adventurous few had sailed along the west coast of Africa as far as the tropics, or northwards to southern Britain.

In the following centuries, which witnessed Alexander's great campaigns, the Trans-Caspian lands, Persia, India and much of north-east Africa became at least vaguely known. Thence to the 14th century progress was proportionately less rapid. The exploration of the Arabs revealed the African coasts in the west, southwards nearly to the Equator, and in the east as far south as Madagascar. The largest accession to geographical knowledge followed the travels and writings of Marco Polo which revealed in outline much of Asia, known hitherto rather dimly through trade contacts since Græco-Roman times. But although the shape and size of the earth as a whole were already known to scholars of ancient Greece, a great part of it long remained undiscovered.

The rounding of the Cape and the discovery of the Americas in the late 15th century opened new horizons of unexampled magnitude. The 17th and 18th centuries saw the effective discovery of the inner parts of the Americas, of Australia apart from its dead heartland, and of much of north-east Asia. Towards the end of the 19th century the remaining real *terra incognita* was reduced to the interiors of Africa and North America, and to the Polar regions. The famous fields of exploration of our own day are too well known to need description here: the traverse by land of the Antarctic continent remains the only major challenge to contemporary explorers.

It is evident that the progress of exploration and discovery is similar in kind, if not in degree, to the local "regional survey" of primitive communities; it leads to an ever growing awareness of environment. Navigation and land-survey have grown concurrently with the widening geographical horizons, while at the same time they alone have rendered it possible. For these reasons it has been common in the past, nor is the usage yet quite dead, to regard the geographer as essentially an explorer.

The Royal Geographical Society, founded in 1830, has always been closely identified with this aspect of geography and in the first half-century of its life it very actively promoted African exploration, just as in later years it lent its technical and financial support to Polar and mountaineering expeditions. Its well-known volume *Hints to Travellers* is designed to assist the wandering fraternity of Service personnel and others in making useful observations in unfrequented ground and thus to assist in accurately describing the face of our planet. But, in our own day, the task of primary exploration, the determination of at least the approximate location of the leading geographical features, is wellnigh complete. Romantic interest may still be evoked by the names of the supposed last undiscovered desert oases such as Zerzura and, if we are untravelled and "un-map-minded," we may cling to the illusion of the novelist's Shangri-La or yearn to follow Ulysses into the unknown. But if the geographer be primarily an explorer it is clear that most of his prospect and horizon is gone. Although much useful detail remains to be filled in, the game of *primary* exploration is wellnigh played out.

Another simple but erroneous view would regard the geographer chiefly as a surveyor. If he may no longer explore, may he not map what is already explored? Yet even to this great task there is a rapidly approaching end and one that will be hastened by the methods of aerial survey. It could not avail the geographer long as a major preoccupation. No doubt, as a user of maps, the geographer requires to be conversant with methods of survey, if only to realize the properties and limitations of his essential tool. But in the fully developed art of surveying as practised by the geodesist and the civil engineer he is generally neither competent nor interested.

Let us therefore turn to that second expansion of geographical horizons, concerned not with their superficial extent but with their depth and complexity. This we owe mainly to the development of the natural sciences. If salient dates be sought to mark the progress of events, they might well be the years 1830 and 1859. The former, in which the Royal Geographical Society was founded, saw the publication of Lyell's *Principles of Geology*. The foundations of stratigraphical geology—the succession of strata and their fossil contents, had been earlier laid, but in Lyell's *Principles* was found the first satisfying philosophy of the subject, a true "Theory of the Earth."

Lyell, like his great predecessor Hutton, could find in the geological history of our planet "no vestige of a beginning; no prospect of an end." Its surface had been endlessly sculptured by agents, similar both in kind and degree to those now operative. Its mountains were but temporary successors to others, sometimes even vaster, of a vanished past and its plains repeated the pattern and the processes of predecessors without number. In a word, it appeared that an understanding of the physical geography of the earth required more than description, however accurate and minute. The surface itself had evolved and its antecedents were stamped on its present face.

Lyell's *Principles* was the veritable forerunner of Darwin's *Origin of Species* (1850) which gave order and coherence to the biological science of later years and changed the whole mental climate of mankind. In the geographical field it gave the germ of the powerful and illuminating principle, that the flora and fauna of the earth, and man himself, were significantly adapted to their physical environments—united with them causally and indissolubly.

We might go on to trace the concurrent rise of other specialist studies dealing with life and its environment—meteorology, anthropology, history and the other social sciences. It will serve our purpose better, however, to view the whole great mental awakening as a single complex surge of mind, an essentially unitary process. From it arose the need for "modern geography" as now understood, but also those doubts and difficulties concerning content and method which have periodically disturbed geographers ever since. If we are to trace the development of the distinctive geographical point of view, we must look rather more closely at the historical relation of geography and the natural sciences.

The work of Lyell and Darwin represented the culmination of research, reflection and debate which had been in progress before their day. In the later years of the 18th century the emergence of a well-ordered world picture, with man as an integral element, was foreshadowed. It was at this time that two remarkable men—Humboldt and Ritter—first expressed in their writings the essential spirit, and perceived the necessary future service, of geography. It is important to notice that their major work was begun and largely carried out before the Lyell-Darwin era; for both died in 1859.

Humboldt travelled in South and Central America from 1799 to 1804 and his results were largely worked up during the following twenty years. His great work *Cosmos*, though published in the middle years of the century (1845–62), was outlined already in lectures delivered at Berlin before the *Principles of Geology* appeared. Similarly, Ritter's mind was formed and his geographical equipment gained in the last decades of the 18th century, although the first volume of his famous *Erdkunde* was published only in 1817. He became the first university professor of geography (at Berlin) in 1820.

Recent years have seen a greatly revived interest in the work of Humboldt and Ritter. It has long been usual to pay at least lip-service to their pioneer rôle, but as the writings of both were voluminous, and Ritter's at least not always easy to interpret, it is to be feared that all too few geographers really read them. In seeking to re-orient the rather wandering latter-day course of their subject, however, geographers have been impelled to look back along the century's path of travel and to re-examine the outlook of the great founders.

Much has been written on the difference in training, temperament, method and philosophy of the two men, and their relations, as contemporaries living for long years in the same city, have come under critical review. To this we need only refer here, for in the perspective of time, the convergence of their thought outweighs divergence. Both were engrossed in the coherent relations of the physical and biological phenomena of the earth's surface, and both were clearly aware that human phenomena showed correlation with physical conditions and were an integral and harmonious part of the complete picture, whether on the smaller scale of the world as a whole, or the larger scale of the continents and their regions. A recurrent key-word in the writings of both of them is *Zusammenhang*—literally "hanging together," which we may translate as "context." Their aim was to see natural (including human) phenomena in their natural groupings or contexts with a view to perceiving the causal relations between them. In such a view there certainly lies the root principle of geography as we know it today.

It is by no means an easy principle to ensue and pitfalls both of method and philosophy need to be avoided. It is especially important to notice that geography made, in a sense, a false or premature start. Humboldt and Ritter were in advance of their time and had not the data for the fulfilment of their great projects. It was their successors who perforce faced the difficulties created by the rapid and successful advance of the specialist sciences. To use a metaphor, itself geographical, to illuminate the matter, we may say that the intellectual vanguard of mankind, breasting for the first time the slopes of a Pisgah eminence, glimpsed a synoptic view of the panorama of natural knowledge, and perceived its oneness or wholeness. But as they descended the slope again to enter the promised land, they inevitably split into separate parties following separate tortuous and divergent valleys, each one seeing, with increasing clarity and definition, one aspect alone of the complete landscape. Such divergence started within the life-times of Humboldt and Ritter and has continued since along the by-roads of narrow specialization, until, in our own day, its perils have begun to be perceived and new syntheses in many fields of knowledge are imperatively sought. Geography aims at one such synthesis and the times are perhaps riper for such attempts than at any stage since Humboldt and Ritter wrote.

We are not of course intending to suggest that the specialization was premature; on the contrary it was natural and necessary and, without the great body of facts produced thereby, no satisfactory geographical synthesis could now be hoped for. Primary exploration itself was unfinished when Humboldt and Ritter wrote; the interior of Africa was not in any sense known till 1880, while much of Asia, upon which both pioneers wrote, remained in essence a land of mystery, until the great Russian explorers, together with Sven Hedin, Aurel Stein and Richthofen, opened it up in the latter part of the 19th century.

Perhaps even more important for the future of both geology and geography was the early work of the American Geological Survey—the explorations of Hayden and Powell in the arid west of the United States. Primary exploration, albeit fully scientific in method, was represented also by the *Challenger* expedition of 1870–73, which first charted the depths of the oceans and systematically examined the characteristics and movements of their surface waters. This expedition also provided the data for the first adequate outline survey of the climate of the oceans. Humboldt had perceived the value of isotherms and had used them so far as the then scanty data permitted, but it was Buchan, meteorologist to the *Challenger* expedition, who drew the first world maps of temperature, pressure, etc., the ancestors of the familiar maps of the modern atlas.

Equally important for the ultimate purposes of geography was the publication by Government surveys of large-scale topographical and geological maps. Both France and Britain were covered by adequate topographical maps by 1870 and the primary geological survey of both countries was well advanced. Parallel developments occurred in other civilized countries. The rise of the meteorological services during the latter part of the century marked the birth of synoptic meteorology and detailed local climatology. The mounting tale of official statistics on population, production and trade added further material for prospective geographical synthesis to an almost embarrassing degree.

It is in the light of this vastly augmented body of knowledge that the work of Humboldt and Ritter appears as inevitably under-equipped and the task of their successors almost impossibly increased in scope and difficulty. Mankind was showing a powerful and persistent interest in its environment and becoming in manifold ways more aware of it. But the method of progress appeared to lie in that same "division of labour" so loudly praised by the economists and so manifestly fruitful in material production. Was it then possible for any man or group of men, in Ritter's words, "to sum up and organize into a beautiful unity all that we know of the globe?" To many the attempt seemed over-ambitious, if not chimerical: the burden of Atlas in modern form.

Nor must we lose sight of another great change which had come over the geographical scene during the "wonderful century." During these years man moved, as it were, to his last destinations and habitations on the earth, and with rapidly increasing numbers fully exploited and "developed" its surface for the first time. We live veritably at the end of an age which the contemporaries of Humboldt and Ritter could hardly foresee.

The four centuries from the 16th to the 19th saw greater migrations of mankind than any since Neolithic times, and not the least spectacular transformations of population and culture were reserved for the latter part of the period. Since Humboldt and Ritter first met as young men, the population of the world has more than doubled and the whole vast range of its productive resources has been brought into use by long-continued trial and error. There is now little more "new land" to be occupied.

The rapid westward migration of the American "frontier," culminating in our own day with the occupation of the High Plains and the furthest prairie fringes, marked one of the last major stages of colonization. The process of settlement and economic development of Soviet Siberia, now in progress, marks another.

Although the population of the world shows marked inequalities of distribution, there is reason to believe that it presents a relatively static picture, in that people, economically speaking, are where they belong. We know to a first approximation the world distributions of metals, fuels and other sources of power. We have experimented widely with the available crops and have even tried new ones. Oriental food crops and other agricultural raw materials have been tried out sporadically in the Old World and the New. The latter has offered its distinctive crops, such as maize, manioc and potatoes, for profitable exploitation in the Old World, and, most important of all, the characteristic wheat culture of Europe has implanted itself throughout the temperate grasslands, with resulting widespread transformation of occupancy and economy and the provision of vast stores of food roughly commensurate with the greatly increased world population. In a word, the surface of the earth has been "developed" by the omnipresent human reagent; its patterns have been brought out as are those of the photographic plate by the developing solution.

All this has vital significance for the rôle of geography. It has been possible in all ages and will always be possible to attempt a compendious summary of the *physique* of our planet in terms of the scientific knowledge of the day. But this is the easier part of the geographical task. The great pioneers were deeply convinced that, as Ritter put it, "the Earth and its inhabitants stand in the closest mutual relations," and there is universal agreement that geographical work focuses on the study of this relationship. Yet the pioneers worked in a world of which the *human* development was relative to that of today, immature. Their general technique is still valid in such sparsely inhabited regions as they originally studied.

Although Ritter wrote on Europe as early as 1804, the prime objects of his later detailed study were Africa and Asia.

Humboldt's most significant geographical writing, apart from his "physiographic" studies, concerned Tropical America and, as he himself admitted, the human elements in the landscape were there "recessive." Neither writer could have fully explored the difficulties of applying the geographical method, for instance, to any one of the forty or fifty giant cities of a million or more inhabitants which have arisen since his day. These men founded geography, but necessarily left many of its essential problems unsolved.

CHAPTER II

THE PHILOSOPHY AND PURPOSE OF GEOGRAPHY

Of course the first thing to do was to make a grand survey of the country she was going to travel through. "It's something very like learning geography," thought Alice, as she stood on tip-toe in hopes of being able to see a little further. "Principal rivers—there *are* none. Principal mountains—I'm the only one . . . Principal towns—why, what *are* those creatures, making honey down there?"

LEWIS CARROLL, *Through the Looking-Glass*

WE have indicated certain difficulties and complications in the more recent growth stages of geography to explain, if not to excuse, the attitude of the rest of the learned world to this subject and to show the measure of plausibility enjoyed by the attacks of its critics. The reader will not expect professing geographers to quake unduly before the admitted difficulties of their task or to accept meekly the strictures on their subject made by others. It none the less remains true at present that any explanation of the subject is perforce a defence of it.

It is no doubt inevitable that persons of a severely analytical turn of mind who have spent the best years of their lives in the careful and minute study of the life-habits of fungi or the chemistry of the alkaloids should look askance, at what seems to them to be the overweening and preposterous claims of a synoptic subject which attempts a simultaneous view of an extensive field. Literary scholars who dig deeply in narrow plots of classics or of history are often no less repelled. But this is not to say that geography lacks a definitive aim and a recognizable scope. Indeed its *raison d'être* and intellectual attraction arise in large part from the shortcomings of the un-co-ordinated intellectual world bequeathed us by the specialists.

Only two questions really arise concerning the validity and worth of geography. Is its programme philosophically rational and desirable and does the application of its methods lead to results interesting and practically useful in themselves?

There is no dearth of adequately formulated answers to the first question. The German and French geographers of the latter part of the 19th century, notably in advance of their colleagues elsewhere (where such existed) in theoretical grasp, wrestled long and earnestly with the problems arising from the Humboldt-Ritter concept. None has stated the general position better than the veteran German geographer Hettner when he wrote:[1]

> "Reality is simultaneously a three dimensional space which we must examine from three different points of view in order to comprehend the whole. From one point of view we see the relations of similar things, from the second the development in time and from the third the arrangement and division in space. Reality as a whole cannot be encompassed entirely in the systematic sciences, sciences defined by the objects they study. Other writers have effectively based the justification for the historical sciences on the necessity of a special conception of development in time. But this leaves science still two-dimensional; we do not perceive it completely unless we consider it also from the third point of view, the division and arrangement in space."

Or as Kraft even more concisely expressed it:

> "Stones, plants, animals and man, in themselves objects of their own sciences, constitute objects in the sphere of geography in so far as they are of importance for, or characteristic of, the nature of the earth's surface."

Both these passages embody essentially the Humboldt-Ritter concept and they might be matched by dozens of others

[1] Cited by R. Hartshorne, "The Nature of Geography," *Annals of the Association of American Geographers*, vol. xxix, nos. 3 and 4 (1939)

made by the masters themselves and their successors. In such terms the philosophical position of geography can be clarified and its aims defined; but we must note two further points if these statements are to serve our purpose. First, they imply a geographical *method* of seeing things together in their spatial relationships which is available and indeed necessary in other subjects. It is habitually used in certain branches of geology and is obviously applicable in the field of plant geography to which Humboldt himself made notable contributions. Geographers cannot pre-empt an obviously valid method for their exclusive use. Geologists, botanists and others will inevitably employ it. It might seem indeed that the synoptic method of examining the *Zusammenhang* of associated non-human phenomena in space was common to the field sciences and that the geographer as such is not required. But in fact analytical sciences are not generally conspicuous for the facility and success with which they use this method.

The second point is even more vital. It is in bridging the gap between physical and human phenomena that geography finds its distinctive rôle. The complications liable to ensue are manifest. Human phenomena involve the activity of mind and purpose. They are, in an important sense, generically different both from the inanimate features and even those belonging to the lower forms of life. At the worst the door seems open to an extreme formlessness here. Must we in addition to our acquaintance with the physical and biological elements of the environment compass human history in its entirety, together with anthropology, sociology, psychology and other social sciences? The matter here at issue is so important and confusion about it so liable to arise that we must consider it carefully.

In broad terms it is evident that geography concerns Land [1] and Man. The field can therefore be approached from the side of either Land or Man and it is unprofitable to debate which is the better approach. Providing indeed that the final viewpoint comprehends both there is little to choose between them. It has been common for some geologists and historians to become

[1] The term "Land" is here used in a wide sense to mean the physique of the surface of the earth in all its aspects, land, air and ocean.

geographers; indeed before the subject emerged as a formal branch of knowledge it could recruit its students only from those trained in other fields.

But there is a danger of a false dichotomy here—a division of the subject into "physical geography" and "human geography." Such cleavage is the very thing geography exists to bridge, and it is false to its central aim whenever and for whatever reason it recognizes or emphasizes two "sides" in the subject. In such a matter as this, there is risk of becoming lost in a mere maze of words, and from the charge of becoming so lost, geographers cannot be entirely acquitted. But recent years have shown a notable convergence of thought such, as may be illustrated by the following statements. "Geography," wrote Vidal de la Blache, the late leader of the French school of geography, "is the science of places not of men; it is interested in the events of history in so far as these bring to work and to light in the countries where they take place, qualities and potentialities which without them would remain latent." With a curiously similar and salutary emphasis the American geographers Sauer and Leighly wrote: "Geography has never been a science of man, but the science of the 'land,' of the earth's surface." These are the opinions of wholly independent writers and are drawn from different contexts. Yet both come very near the heart of the matter and illumine our problem greatly.

There can be no doubt that what interests the professional geographer and the layman, as geographer, on their travels is the essential pattern and quality of the earth's surface—"places" or "areas" and the great difference between them. In its simplest essence the geographical problem is how and why does one part of the earth's surface differ from another. To describe and comprehend the earth's surface certainly requires that we lay the natural sciences under contribution, but in some respects the greatest differentiating agent is man himself. It is he who makes and maintains the difference between town and country, between "the steppe and the sown." In ways innumerable, man's effort has developed the face of our planet, emphasizing natural differences and bringing into being others which did not exist before his advent.

All this is not to say however that the *primacy* of Land and Man is equal. In one sense it is profoundly true that the most important part of geography is the "human part" since without it the subject would lose its major rôle. But it is equally true that the foundations of the subject lie on the physical side and, as with a building, foundations come first, however vast the superstructure. This may sound the veriest truism, yet it needs emphasis.

It is not uncommon to find teachers and students of geography who lament that they are "weak in physical" as if this were a regrettable defect in their constitution like a weak backhand at tennis or a tendency to "slice" at golf. The implication is that they can be depended upon for a notably better performance on the human side. It is just this differentiation between sides which geography cannot recognize. The serious study of the subject cannot begin without the findings of physical geography; the play cannot proceed without a stage and it is a stage, be it noted, which plays a much larger part in the action than in theatrical performances.

The tendency of this discussion may seem to shape unfairly against those who claim the title of "humanists." We may be reminded in the words of Pope that "the proper study of mankind is man." To this there are a number of evident replies. If we want to study man as such in one of his many aspects, we are afforded a choice not only of history but of the developing group of social sciences. There is no point in taking the ge- out of geography merely to facilitate trespassing in these fields which exist and are studied in their own right. In any case we may be permitted to recall that the point of Pope's couplet was an admonition to know ourselves and not to presume to scan God. It was not intended as a deterrent from the study of Land.

Here indeed we reach the clue we are seeking. In our *first* enquiries in geography we do, to a first approximation, know Man—i.e., ourselves. In these first stages of analysis at least, the human features of the earth's surface appear by no means so complex or so puzzling as the physical elements. This is indeed an illusion which further study corrects, yet it is true that we do not need formal elementary instruction in the facts that man clears forests, builds dwellings, cultivates fields or tends herds, works mines and factories and forges links of communication. It is the nature of man so to do. This statement would no doubt evoke indignant qualifications from anthropologists. But so far as concerns the *elements* of geography it is futile to assert that "human" or "social" geography can be seen in terms of formal categories and universal principles and processes as can physical geography. This imputes to it no inferiority; it is rather to admit that it is infinitely more complex, subtler, more flexible and manifold.

There have been, it is true, some attempts to formulate the human bases of geography in terms comparable with those of physical geography. They are generally far from convincing and often almost ridiculous. The most successful treatment is probably that of Jean Brunhes. We cannot find equal satisfaction in certain American attempts to treat this field. Thus Finch and Trewartha, in a most useful introductory work *The Elements of Geography*, devote nearly 560 pages to "the natural elements of landscape," followed by 30 on "the cultural elements of landscape"—i.e., the elements resulting from man's use of the land. This disproportion is startling and is by no means wholly due to the lack of development of "cultural anthropology." The limitations of the approach adopted may be judged from such statements as these. "The American farmstead is composed of the residence, which is the nucleus, usually housing a single family together with a collection of various barns and sheds of various sizes and shapes used for housing livestock and for storing crops and machinery. A fence commonly encloses the farmstead and separates it from adjoining fields."

While it is of course true that not all the farms of the world are precisely of this pattern, the statement is so general and commonplace as to lack any great measure of interest or usefulness. Similar criticism may be made of the following under the general heading "Distinguishing Features of a City." "Within the down-town business district tall closely spaced substantial buildings of brick, stone and concrete occupied by retail shops and professional offices prevail. This is emphatically the hub or nucleus. Upon it the street system converges so that within it the traffic is usually congested." It is evident that within the confines of time and space represented by "Western culture" such statements as these are unbearably trite; they may be taken for granted. It is quite impossible to manufacture a human basis for geography in this way, comparable with the physical basis—or if it be not impossible there are few signs yet of its being successfully done.

We necessarily grant, of course, that both the form and function of the features of the cultural landscape are found to change more radically if we widen the area of reference and, taking the world as a whole, some knowledge of cultural anthropology is necessary to interpret what we see. Professor Daryll Forde's work, *Habitat, Economy, and Society* may be consulted in this regard as a necessary corrective to the narrower forms of geographical thinking. But even here we must note that he is dealing for the most part with the economies of primitive peoples which, in dwindling numbers, occupy a small and decreasing fraction of the area of our planet. Such knowledge may deepen the geographer's philosophical grasp but it does not greatly extend the range of his work-a-day tools. Man is, after all, a single interbreeding species, and despite all cultural divergences and variant stages of economy, his basic needs for food, shelter and community life show at least strong "family" resemblances throughout.

We must necessarily be at some pains to emphasize that it is just this and nothing more that we are seeking to maintain in respect of the human foundations of our subject. That geography is itself to be ranked among the humanities and that knowledge of human economies and their evolution are necessary to the geographer are propositions which no student of the subject could deny. But in the preliminary stages of enquiry we can make few *general* statements about man and his work in the large which are both true and worth making; there is as yet no unitary and compendious "science of man."

If we were concerned to explain the face of our planet to an intelligent non-human visitant from some other sphere, we should no doubt have to attempt some such general statement to render our geographical report in the least meaningful, but the "human" student of "human" geography is in different case. For him we cannot blend borrowings from history and the social sciences so as to create a generalized human geography. The nature of man and of human evolution ensures that each region which we study is in large measure unique. There may indeed be analogies and parallels between them; but no prior knowledge of the human and historical background of one will necessarily avail us in any other. Ideally at least we cannot know too much of the history of any region or area whose geography we seek to interpret. But, to take an instance, the writ of English economic history does not run in China or Chile; the plot is different for each geographic theatre, and each must be studied on its merits. In terms of physique the Argentine pampa recalls the Canadian prairie, but in the human development of the two the differences are on the whole more important than the similarities.

It is this uniqueness of the geographic region as humanly developed which has seemed to some to deny to geography the status of a science in the narrow sense. One cannot in any real sense deduce the human resultants from the physical causes. Attempts at such deductions have of course been made. The German geographer Ratzel and his followers are commonly regarded as guilty of the heresy of "determinism." The idea was not a new one, nor is it yet by any means dead. It may be most concisely expressed in the words of the French philosopher, Victor Cousin, when he wrote, "Yes, gentlemen, give me the map of a country, its configuration, its climate, its waters, its winds and all its physical geography; give me its natural productions, its flora, its zoology, and I pledge myself to tell you, *a priori*, what the man of this country will be, and what part this country will play in history, not by accident but of necessity, not at one epoch but at all epochs." Our growing knowledge of geography has long since destroyed the basis of any such exaggerated claims.

The case against determinism has been clearly, almost violently stated by Lucien Febvre. There is even a possibility that the pendulum has swung too far, for geographers have become extremely sensitive to the charge of "determinism." There is no need to deny that "Land" exerts direct influences on "Man," but every need of extreme caution in investigating them. Our knowledge is too slight and the time range of our observation far too limited to justify facile hypotheses on the subject. We have here again a subject apt to produce exasperating and quite unsatisfying play on words. Some have sought refuge in the comfortable formula that the reactions of Man and Land are mutual. This is evidently true without being either profound or helpful. We can however take it as a wise practical rule, that, in the first instance, we are well employed in studying the imprint of Man on Land—what some have called the "cultural landscape." It will be time enough to seek to isolate the effect of Land on Man after much further detailed investigation.

Let us note however that in a tacitly limited or qualified sense an element of determinism remains inevitable in geography, at least in its modes of expression, and it is quite unhelpful to express exaggerated disgust at some of its accepted forms of statement. Thus if, in a geographical context, we are asked to "describe and account for" the distribution of population in Sweden, our "accounting for" cannot and will not include the myriad individual decisions which entered, unrecorded and unexplained, into the making of the settlement pattern.

We still, in a sense, account for what we describe in pointing out that the population groups itself in accord with economic potentialities, within the range of our present resources. We do not assert that the population has never been grouped differently or that its pattern cannot change. We are content with noting such facts as that the upper edge of continuous occupancy coincides roughly with an "altitudinal limit," that major urban population groups are related to facilities for access by land and sea, and many similar correlations. In this way we certainly "account for" the distribution of population without begging any philosophical questions. All that is implied is that in a local and temporary sense man disposes of himself and his structures and fabrications in sites and situations which are rational, or which at least appear to him to be so. A secure and prosperous farming community will attach itself to land favouring arable cultivation, and will initially at least avoid marshes as unprofitable and negative ground. We follow their natural thoughts and reactions as we note their distribution of their material culture.

But in other circumstances marshes may give shelter to settlement in disturbed and warlike times. Hereward's Ely stands as symbol of a recurring pattern in the law of men and places. And if the Fenland morasses are today smiling and productive fields, this is but one more illustration of the principle that "geographical values" change with time. But in each phase the human, re-acting upon the physical circumstances, may be fairly said to "determine" geographical pattern. It is only those geographers who choose to assume the rôle of amateur word splitters who will trip persistently in traps of their own making concerned with the word "determine."

We must pass now to another general principle of importance to our geographical thinking—primarily a question of scale. Geography may be pursued by two methods or upon two levels, which we may distinguish as General (or World) Geography and Special (Regional) Geography. This distinction was perhaps first clearly made by Bernard Varenius (1622–1650) in his *Geographia Generalis*, but its reality is apparent to every modern student. On the one hand, we can make certain generalizations about world distributions both physical and human. There is a world pattern of land and water and a world pattern of climatic or plant regions. With equal propriety we can deal with the world distributions of population, religions or language. Here as always the geographer is seeking "Zusammenhang."

It is not necessary that the generalization shall concern the whole world. What is characteristic is the *comparative* method applied at least over some large area. Thus Vidal de la Blache in a well-known map (Fig. 1) shows nearly fifty famous towns of the Old World lying on or near desert borders. Though they vary greatly in latitude, climate, geological setting, local site, and human context and background, they show a significant relation to the edge of the arid lands. A generalization of like

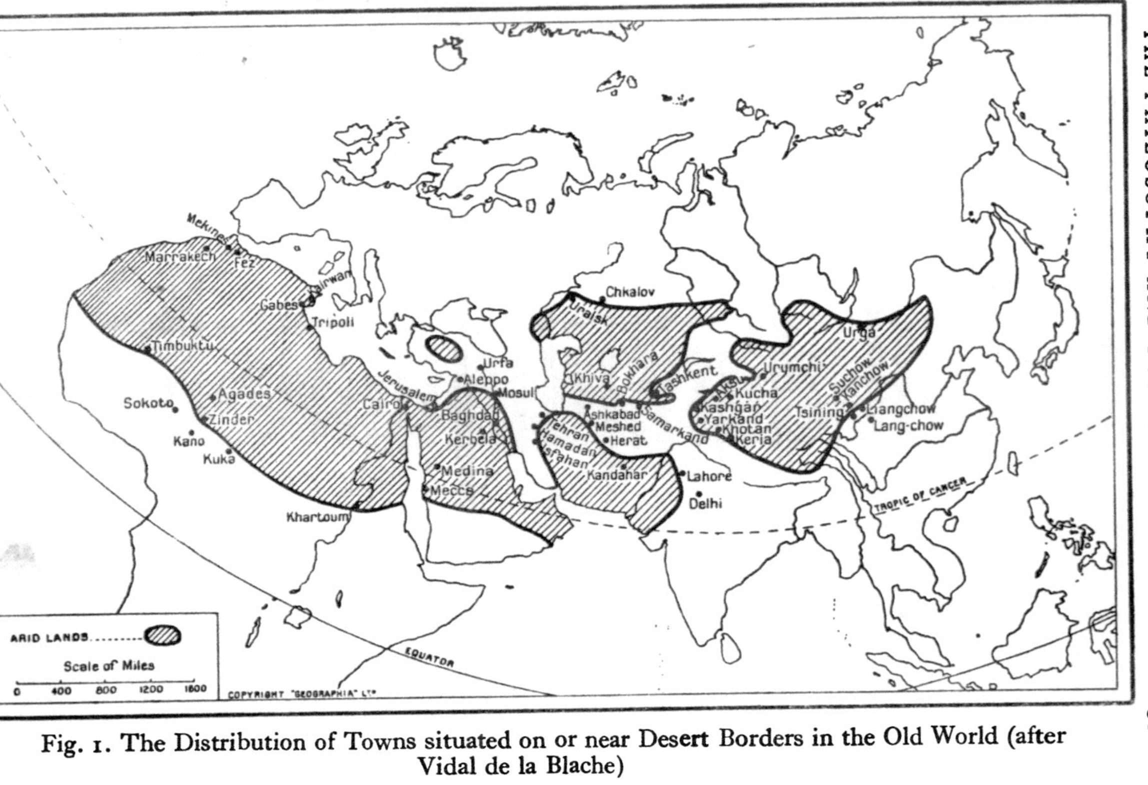

Fig. 1. The Distribution of Towns situated on or near Desert Borders in the Old World (after Vidal de la Blache)

type and order was made by Vaughan Cornish in pointing out that at the end of the first century A.D. the northern limit of the ancient cities of the Old World between Holland and Korea coincided closely with the annual isotherm 48.5°F. Again we might take, with Professor C. B. Fawcett, a wide and comparative view of the peculiar distribution throughout the world of cities with more than a million inhabitants (Fig. 2). In all these cases we are engaged in the selection and comparison of facts over wide areas, portrayed on maps of small scale.

By contrast a day's journey in the Weald of Southern England reveals an order and regularity in the landscape which must be studied on the ground or analysed on large-scale maps. Here, with a perfection of detail rarely equalled, the human landscape is related to land-forms and geological outcrops. It is such regions that Professor Rodwell Jones had in mind when he wrote: "Man's activities, his points of settlement, lines of movement, occupation and economic development bear quite special relation to some few and essentially simple physical characteristics." Yet the geography of the Weald is "special" or "regional" geography in that the detailed relations it exhibits are unique. Few generalizations drawn from it would be found valid in another area, and none so derived would have a world-wide relevance.

We have only to cross the narrow seas to the Boulonnais of France to find a region almost identical in physical constitution which, because it falls within the boundaries of another state and thus shares another social and economic environment, presents a very different "cultural landscape." Each "region" is manifestly a unique part of the earth's surface with a unique history of development both physical and human. Many of the "repeating patterns" which interest the geographer occur *within* such a region; they do not extend in any large measure outside it.

Here then are two very different levels of geographical study. World or General Geography is *Geography* in the literal sense. Regional or Special Geography proceeds with the same general methods and aims, but in any one study it is concerned with much less than the earth, except in so far as the part is

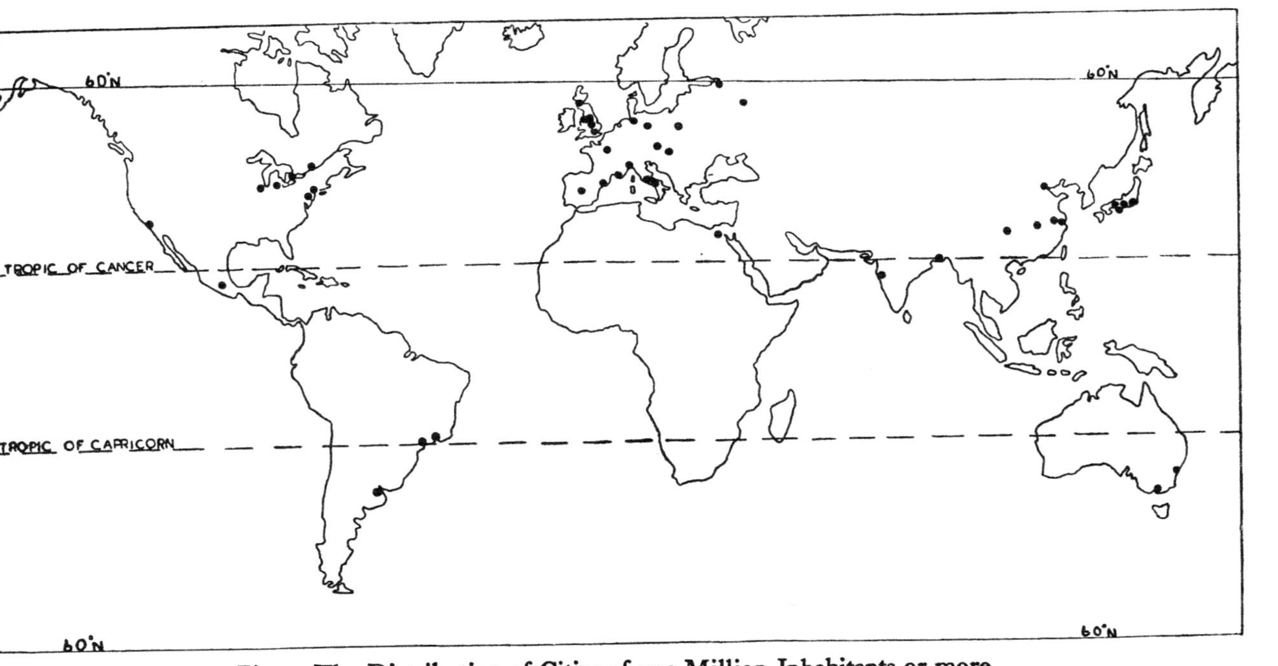

Fig. 2. The Distribution of Cities of one Million Inhabitants or more.

always influenced by the whole. We return to this point in a later chapter; for the moment it is enough to note these two great sub-divisions of the subject. General Geography, as we shall see, involves a number of distinct and systematic studies. These find their application in the description and interpretation of actual areas or regions, large and small.

CHAPTER III

PHYSICAL GEOGRAPHY AND BIOGEOGRAPHY

The earth is rude, silent, incomprehensible at first, . . .
Be not discouraged, keep on. . . .

WALT WHITMAN.

THE interest and importance of physical geography is not, as a rule, doubted by the non-geographer to whom it seems an obvious and necessary part of the subject. Nor dare the geographer of any school of thought deny its relevance. Yet, academically speaking, it has become something of a cinderella in Britain and America for reasons not difficult to imagine, but impossible to excuse. The necessary scientific background of the subject tends to render it difficult and unfamiliar if not "arid" to the humanist. Since it would plainly be absurd to seek to dispense with it altogether, the line of least resistance for those inclined to such a view is to reduce it to its very simplest terms and to regard it as basic or preliminary training which may be "got through and left behind." Such simple essence may be deemed to comprise the descriptive study of relief and of climate, as represented by monthly mean figures of temperature and rainfall. It requires no great intellectual effort on the part of geographers or non-geographers to compass a physical geography so limited. It is only when we begin to increase the precision and the detail of our geographical picture that physical geography becomes in the best sense "technical" and its study in any sense a discipline.

Physical geography is, in a sense, better organized than its human or social counterpart because it rests upon specialist sciences like geology and meteorology which had made great progress before the aims of modern geography were formulated in any detail. There is thus no dearth, but rather an embarrassing wealth, of material out of which to construct the subject. The "social sciences" have, on the whole, developed less rapidly; it is claimed today that they have been starved of research facilities. In any case their subject matter is far less amenable to precise statement and they are, in general, denied the powerful tool of experimentation. This, as we have seen, tends to retard the growth of a purely human geography. We may take comfort in the fact that even if such growth had been possible, it might well have been premature.

The physical aspects are literally basic to a full geography, and for the good health and future prospect of the subject as a whole, the early advance of physical geography may be rated as a fortunate circumstance. It is none the less fair to concede that, tactically speaking, the geographer often feels some temptation to belittle the physical and magnify the social aspect of geography, since brother-specialists all too readily assume that geography is merely physical geography.

Historians, economists, sociologists and others confidently assume their own competence to deal with the whole field of man and his works. They remain permissibly ignorant of consequent streams, fault-line scarps, föhn winds or podsolic soils and are inclined to think that such, if anything, form and fill the field of geography. In most current projects for "social studies" it is allowed that the geographer may contribute a little on "land-forms" or "climate," but the rest of his field is, too often, complacently and officiously wrested from him. Yet he can hardly hope to reconquer his lost provinces without his major weapon—a full and detailed knowledge of the physical environment.

It is just this knowledge which his brother-specialists lack and in reference to which he can best demonstrate the extent of their own ignorance and the curiously incomplete and erroneous nature of some of their own judgments. They, on their part, will pay little respect to a physical geography so simple and elementary that it is within the competence of any educated man. They rightly demand more than this and the geographer is well advised to see that they get it. All this is plainly to say, therefore, that there is need of the specialist as well as the general development of the field of physical geography.

Physical geography is itself divisible into sections dealing with Land, Air and Ocean respectively. It is a truism to say that the three are intimately linked, but it is equally true that in important respects they are curiously different and separate. Each rests on its own special science—geology, meteorology and oceanography respectively and we may best proceed by examining the relation of physical geography to each of these.

Physical Geography and Geology

"Geology," wrote Sir Archibald Geikie, "is the science which investigates the history of the earth. Its object is to trace the progress of our planet from the earliest beginnings of its separate existence through its various stages of growth down to the present condition of things. Unravelling the complicated processes by which each continent and country has been built up, it traces out the origin of their materials and the successive stages by which these materials have been brought into their present form and position. It thus unfolds a vast series of geographical revolutions that have affected both land and sea all over the face of the globe."

The substance of this authoritative statement has often been differently but rarely better expressed, and it serves to make clear at once how close is the relationship between the two "Earth sciences." Though the geographer's knowledge must be wide, it is safe to say that of all subjects the one of which he must perforce profess some knowledge is geology. Yet to apply here Ritter's oft-quoted dictum, the geographer must take from the geologist, "a portion not the whole" of the latter's science, the portion which is relevant to the surface of the earth and the human environment. Such selection is difficult and, if narrowly conceived, even dangerous to the real well-being of geography. Dangerous because in our anxiety as geographers "to keep our aim simple and clear" we may tend to forget that knowledge is indivisible and its divisions entirely man-made for reasons of mere convenience.

It cannot be doubted that a knowledge of geology as such and in all its branches is not too much to ask as a background to physical geography, and much preferable to a series of *ad hoc* borrowings. None the less many of the specialist investigations of geologists diverge far from the geographer's field and cannot profitably be followed by him. This is true particularly of the fields of palæontology and of mineralogy and crystallography. The structure of geological science is entirely simple and logical, but, as worked out in curricula and courses, it may seem a little curious and even repellent to a beginner with geographical interests. It is a question of ends and means. With its declared "historical" aim as stated by Geikie, it must still recognize that rocks are the documents of earth history and the registers of past geographies, that rocks contain fossils by means of which they can be dated and the conditions of their formation perceived, and that rocks are made of minerals, most of them crystalline in constitution.

Thus it is that many a tyro, attracted by the study of landscapes and their origin and seeking the key to interpret them, is plunged with brutal and rather unimaginative zeal into crystallography. It is a long trail that he here starts upon, but it has an end which will bring him back to the surface of the earth with its varied form and constitution.

All that we need say here is that the student of geography if not the better is at least none the worse for a prior training in the basic geological disciplines. Yet it is admittedly possible and sometimes preferable to enter the geologist's "building" and explore its upper floors otherwise than by the basement. It is possible to have a perfectly satisfactory and intelligent comprehension of a sandstone in its contribution to past geological history and to present geography, without reference to the crystallography of quartz, its chief mineral. The geologists of former generations did not doubt this, for they themselves were denied the laboratory facilities enjoyed by beginners today.

A much more fundamental division of interest between geologist and geographer lies in the fact that geology is essentially historical in its outlook. Geologists think of the sequence of events and phenomena in time and they necessarily tend to look backwards in time. This is a valuable mental habit from which the geographer should not seek to absolve himself, either in the physical or the human fields, providing it does not obscure his sense of the present. An example may make this clear. Throughout the Western Weald, in Sussex and Hampshire, the formation known as the Upper Greensand forms a bench terminating in a minor escarpment at the foot of the Chalk downs (Fig. 3). Bench and escarpment are geographical features which locally dominate the landscape. The formation gives rise to distinctive soils of high arable quality, springs break out where it rests on the underlying Gault clay, and villages cluster on or near its outcrop.

But, in quite another aspect, the Upper Greensand is a sandy marine deposit containing fossils. To examine the rock and its fossils, to trace it eastwards until it is replaced by clays (Upper Gault) or westwards where it finally reaches the vicinity of the former shoreline in Devonshire is to collect evidence for a reconstruction of the geological past—a piece of palæogeography. The geologist may be primarily concerned with this, or alternatively, and with the same ultimate aim, with the detailed study of the fossil contents of the deposit and its mineral constitution. His work in these fields will contribute little or nothing to our comprehension of the Upper Greensand as a landscape maker, a parent material for soils, and a reservoir of underground water. Such represent the "portion" rather than the whole of geology which geographers, are adjured to take and use.

Aside from the value of geology as a general geographical background, it is clear that the common ground of the two subjects includes at least the study of land-forms—geomorphology properly so-called. The land-forms of an area reflect the "lithology" and the structure or attitude of its constituent rocks, but they reflect also the work of general atmospheric destruction or "weathering" and the erosive activity of rivers, waves, wind or ice.

One gains one stage in the understanding of the earth's surface in realizing that it has been literally sculptured by these agents. That streams, in general, make the valleys in which they flow, that cliffs recede under the attacks of waves, that ice and snow shatter and quarry the face of precipice or peak, all this is a commonplace of both physical geology and physical geography

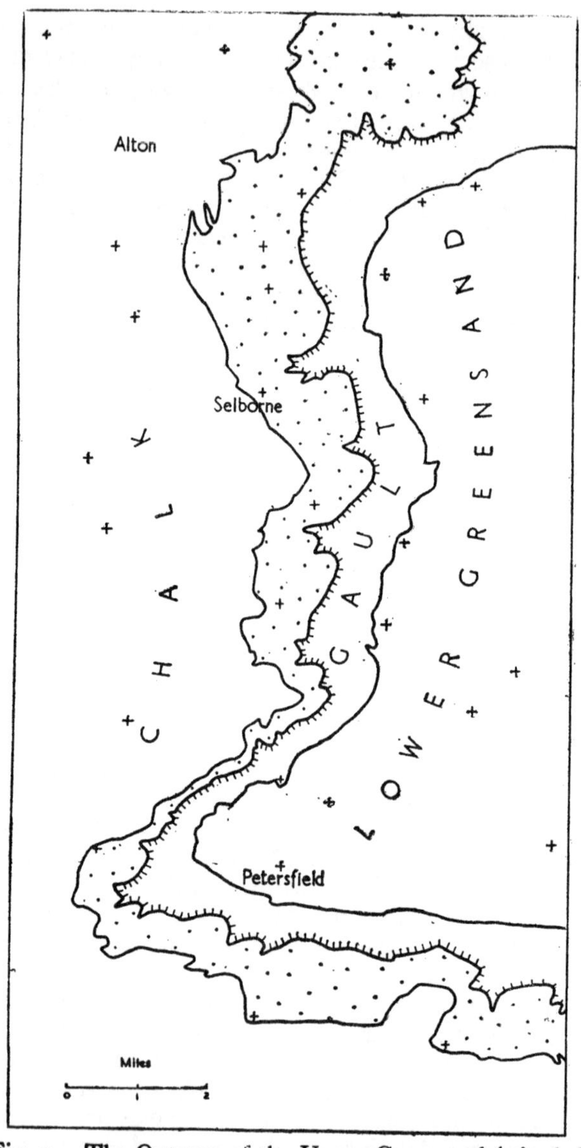

Fig. 3. The Outcrop of the Upper Greensand (stippled) in the Western Weald and the Villages sited upon it.

between which there is little distinction or difference. We can get so far knowing no more of geology than this so-called physical or dynamical geology, but so far is not far enough. Episodes far more ancient than the vicissitudes of sculpture are present in most landscapes; compression, distortion, upheaval and subsidence have contributed their important and sometimes dominant quota to the visible landscape.

No good purpose whatever is served by unedifying debate as to whether the study of the origin of land-forms is really geology or geography. It is necessarily both. Like bio-chemistry or astro-physics it cannot be claimed wholly by either of the partners or parents between which it stands. Logically, the processes of origin of the landscapes form the last paragraph of the last chapter of the long tale of geology. But these same scenic features are literally basic to the geographer's patterns. The only matter which has ever been at issue is one perennial in geography and its borderlands, the question of "time versus space." How far is it necessary for geographers to study the origin of land-forms? Why not accept them as "given" on the ground that it is their form, constitution and relations which matter to us, not the long series of which they are—to date—the last term.

Here again we confront the geographer's dangerous choice of what to accept and what to reject in composing his subject and organizing his training. There is some ground for scruple concerning the relevance of land-form origins. For example, the sharply bounded plateau of the Blackheath Beds, which overlooks Thames-side between Deptford and Woolwich, was associated on the older geological maps with a line of faulting, leading one to conclude that the feature was a "fault line scarp." Further evidence has abolished the fault from the newer maps and we should now conclude that the feature was simply and solely river-cut. Yet the influence of this "plateau brink" on the site and growth of the river-side settlements and indeed upon the whole geographical lay-out stands unaffected by either hypothesis of origin or any other that might be offered. Such an argument, though it may seem unanswerable, must, in fact, be rejected on various grounds.

The best classifications, including those of land-forms, are genetic, i.e., based on genesis. The best way, in general, of understanding anything is to understand how it has evolved or developed. It is not only in geomorphology that this principle works; it is equally relevant in historical geography. To treat geography too literally as an affair of the "quasi-static present" is to make both it and its students seem foolish and superficial. It is true that our primary aim is to describe the present landscape; but it is also to interpret it. This cannot be done in ignorance or forgetfulness of the endless flux, not only of the physical phenomena but also of the life of human societies.

Our study has therefore always to be evolutionary, at least in the short-range sense. It is unscholarly to take either land-forms or human societies as "given" and static facts, though we must not let temporal sequences obscure spatial patterns. Between this Scylla and this Charybdis, the geographer seeks to steer his careful way and, if on occasion he humanly errs, the intellectually charitable will at least be inclined to forgive him.

Another compelling reason, in Britain at least, for the continued study by geographers of geomorphology is its very common neglect by geologists. Though it is logically within their province as they tacitly acknowledge, there is a certain divergence in method between geomorphology and the other branches of geology. As Geikie wrote nearly half a century ago: 'The rocks and their contents form one subject of study, the history of their present scenery forms another." In harmony with this statement, his *Text-book of Geology* of which the opening lines have been quoted above, devotes only its last twenty-five pages, less than a fiftieth part of the whole, to "Physiographical Geology."

One recalls other works of this great geological writer, notably his *Scenery of Scotland*, which are contributions to geomorphology, but here, as in Lord Avebury's well-known works, the subject of scenery is employed as a popular approach to geology and the scientific discipline of geomorphology, as later developed, hardly figures. Such later development we owe largely to American writers, and in the United States geomorphology almost ranks as a separate subject. Meanwhile, over most of the world, including parts of Britain, though the rock succession and structure have been made clear and the major episodes of past geological history are known, the last and, for the geographer, the most important chapter is perfunctorily written, if at all.

In such circumstances the geographer must be prepared to obtain for himself the materials he needs. If and when the geomorphology of the world is known as well as its stratigraphical geology, it may be time to re-define the attitude of geographers to it. Such time is not yet and, even when it arrives, it will still be necessary for geographers to be fully conversant with the technical nomenclature and methods of investigation of land-form study.

If the mind and the eye of the geomorphologist be the major contribution of geology to geography, it is not the only one. Not land-forms only but structure as such contributes to the human environment in respect of the location of underground water and petroleum, of coal, metallic ores and all other forms of mineral wealth. A major fault line may coincide with a complete transformation in economy and cultural landscape, even though it makes no physiographic feature. Again, geological constitution is a part determinant of soil patterns. The rural landscapes of Southern England or Northern France, to name two examples only, cannot be interpreted without the geological map in hand. It explains so much that the humanist may be tempted to suppose that it explains everything. But it certainly provides the major proximate explanation of the distribution and density of settlement and of the pattern of agricultural geography. It would avail us less directly in regions of more extreme climate such as Russia, or the humid tropics, where soil character is a far less detailed reflex of geological parentage, yet even there it remains significant. In this matter of mineral wealth, water-supply and soils indeed, the geographer must make immediately relevant borrowings from economic geology, since these aspects are directly significant in the study of the earth's surface as developed by human action.

In the final analysis, the geographer cannot fail to become aware that the real vindication of the geological approach to geography and the explanation of the large part played by geological material in general physical geography lie in what may be termed the "principle of continuity." Geology, as W. M. Davis averred, is the geography of the past, and geography the geology of the present. If man as a "geological agent" vastly outvies all other living creation, this truth yet remains quite unaltered.

There is and must always be, a sense of intellectual completeness in seeing the human development of a region against a background which is in no sense a dead structure, but one which has evolved and left signs of its past in its lineaments. For example, the great Aquitaine lowland, lying between the Pyrenees and the *Massif Central* of France was once a gulf of the late Tertiary sea, almost literally an extension of the present Bay of Biscay; it was gradually filled up by rock-waste from the surrounding uplands. This process is still going on by virtue of the heavy river deposition of the Garonne and its tributaries. The physical background in this and many other areas is more than mere background to the human history. It overlaps with and is continued in it. Geology in the fullest sense of the term immensely enriches our view and conception of geography and we must always oppose the efforts of those who, jealous or afraid of the "general earth science," seek to expel it from the geographer's field and to build a ramshackle and irrational "non-geological" physical geography in its place.

Physical Geography and Meteorology

The relation of climatology to meteorology is closely parallel to that of geomorphology to geology. In both cases there is a "geographical" part of the pure science but in neither should it be regarded, because "geographical," as beyond the bounds of the other subject, nor can the geographer avert his ken from the true scientific background.

Almost within living memory meteorology was a little developed and relatively unsuccessful branch of science, comprising little more in its subject matter than the description of its instruments, the observation of the "meteorological elements" and the findings, to date, of weather-forecasting. In those days the fundamental problems of the physics and dynamics of the atmosphere appeared almost hopelessly far from solution. The individual investigator, however well equipped as a physicist, could make no advance without abundant data. The provision of meteorological data requires not only national, but international organization, and lack of this was itself a sufficient cause for the slow advance of the subject in its early days.

Now all is changed; not only is there a skeletal world-network of recording stations, but theory has registered great advances. It is a rather common error and one distressing to the meteorologist, to regard his subject as finding its be-all and end-all in weather forecasting. This is only a part, and in some respects a small part, of the meteorological project which is to examine the atmosphere in all its properties, movements and relations. A glance at the pages of Sir Napier Shaw's *Manual of Meteorology* or at the more recent work of D. W. Brunt will show how definitely meteorology is becoming a branch of applied physics, necessarily speaking in mathematical language.

The advance of theory inevitably has its bearing both on weather-forecasting and climatology. In particular, the emergence of the conception of *air-masses* will be a powerful factor in both fields, and it is expressed in language very familiar to the geographer and ready to his needs. We shall say something more of it later; for the present we may note that it implies permanent or semi-permanent "structures" in the atmosphere, air-masses of different origins and properties, separated by relatively clean-cut "surfaces of discontinuity." It cannot be doubted that this new body of ideas is destined to revolutionize the study of the atmosphere on both its analytical and descriptive sides.

It is, of course, impossible here to give even an inadequate summary of the content of modern meteorology, and it is the less necessary since some knowledge of it has been widely diffused by war-time training, and excellent elementary summaries abound. What we must notice is the natural and inevitable divergence of its matter from the field of the geographer's competence. No doubt it was always interesting to speculate on the mechanism of weather and the physical processes of the atmosphere, but the attraction of this field has been vastly enhanced as our insight has grown and real results have been obtained. Increasingly, physical, dynamical and synoptic research put the seemingly sterile archives of climatology in the background, and it may be difficult to conceive that any competent meteorologist, with his mathematical lance duly burnished, would seek employment or satisfaction in this field. Nevertheless it remains important to geographers, biologists and many others.

We confront here the distinction between "weather" and "climate" which requires careful discussion. The antithesis between the two concepts is a double one. In the first place it may be expressed clearly, if over-simply, by saying that the inter-tropical areas have "climate" but no weather, the extra-tropical, weather but no climate. This is plainly not literally true but it expresses the vital geographical and meteorological fact that the "temperate" zones are highly intemperate in their weather, being the meeting place and battleground of "tropical" and "polar" air-masses from whose inter-action are generated the non-periodic pressure changes of our latitudes, with the long train of travelling "highs" and "lows." But the concept of climate clearly retains its meaning even in these climes; we too experience a march of the seasons with seed-time and harvest. It is tempting perhaps to define climate as "average weather," though this is a crude and inexact way of putting it. Rather let us say that climate, in a "weather-ridden" area is the summation or "integral" of atmospheric conditions over, a period of time. Weather is one phase in a succession of phenomena whose complete cycle constitutes climate.

A more fundamental distinction between meteorology and climatology lies in their interest in causes and results respectively. The march of the seasons in, say, the Sudan, is an exceptionally simple and a vital control of life, including that of man and his economy and social organization. If we express the climate of Salisbury, Rhodesia, in terms of monthly temperatures and rainfalls, we are stating accurate and significant facts from which important consequences flow. But the facts, as stated, throw little light on the physical processes concerned. Geographers have, indeed, been all too ready to assume the contrary.

In tropical climates "rain follows the vertical sun" is the rough correlation which appears to emerge, but the rain falls from actual clouds at specific times. Even though the weather element is less marked and obvious, let us say in Salisbury, Rhodesia, than in London, we cannot say that weather is absent. As in London, the sky changes from hour to hour, during the rainy season at least, and a bald climatic statement would be of little avail to an airman on a flight to Cape Town, when the height and thickness of the clouds and many other measurable and forecastable elements would be of prime concern. The ease with which we have dealt, in past years, with tropical climatology has been the measure of our ignorance of tropical meteorology, a fact thrown into clear relief by the demands of the war-time weather services.

It has, in fact, been true for many years that climatic maps of the world, while giving a description of the facts within the limits of available data, have carried many anomalous and unexplained features. About the turn of the century, when Herbertson and others were first clearly formulating the idea of climatic regions, such anomalies were all too frequent. They less frequently concerned temperature than rainfall.

It is not difficult, qualitatively at least, to rationalize the distribution of recorded surface temperatures in air or ocean, though it would be a formidable task to attempt to calculate the distribution from the known facts of solar radiation. Rainfall anomalies concern even more deeply the physical heart of the subject, the character and limitations of adiabatic expansion of moisture-laden air, and above all the mechanism of the atmospheric circulation. Confronted with such anomalies, the geographer was tempted to rest content with results and to remain ignorant of causes. It is easy to defend such an attitude, but it has its dangers. We are little inclined to rest content with no explanations and sorely tempted to accept wrong or partial ones in lieu of adequate analysis.

Though the facts of climatology will always be of high importance to the geographer, he must remain aware of the trend of meteorological research and be prepared to appreciate its conclusions. The basic principle at stake is as relevant as in the case of geomorphology. Only the genetic approach can give a true picture and it is possible now to conceive of a true dynamical climatology increasingly informed as to the physical background of climatic facts. There is, it is true, some measure of difference between the two cases, for only the occasional specialist student in geography will be equipped for primary research in meteorology, whereas numerous geographers have made and are making contributions to geomorphology. But the gap between climatology and meteorology can be permitted to widen only to the detriment of both, and an elementary basic training in meteorology for geographers cannot be dispensed with. Provided with this, he still faces the main problem, that of "climatic control" in the widest sense. It is one full of pitfalls. Corneille's lines:—

> "I dare to say, my lord, that to all climates
> All kinds of states are not alike adapted."

express a vein of speculation often pursued both before and since, and a small library of books could be found to maintain the thesis that climate makes man and his civilization.

We must face the fact, however, that our reliable knowledge of the direct effects of climate on man is deplorably scanty. We dare not even assert any simple relationship between climate and man's physical characteristics, though theories, plausible if unscientific, have abounded. But this does not justify us in ignoring the palpable control exercised by climate on human economic and social rhythms. Nor must it be permitted to disguise the importance of climatic changes as a factor in history. Both these aspects are central to the interests of geography and if they are suspect by reason of the rather uncritical character of some of the writing thereon, this leaves them none the less fertile fields of research.

Oceanography

That oceanography should play a relatively small part in the general conspectus of physical geography is a natural reflection of the fact that the oceans are, humanly speaking, "deserts." Nevertheless the science of the sea in its own right has grown to extended dimensions. Maury's *Physical Geography of the Sea* (1855) is commonly regarded as the first attempt at a formal presentation of the subject. Its main objective was practical in seeking to bring together information on currents, winds and other matters relevant to navigation, but it shows also a philosophical tone and curiosity about causes which is not hidden by the quaintness and orotundity of the language. Beginning with the famous account of the Gulf Stream and its effect upon climate, it deals also with the general circulation of the oceans and their salinity as well as with many aspects of marine meteorology and climatology. Less than a century old, Maury's pages make clear the paucity of information on which he had to rely, and the difficulty of sub-surface observations.

The progress of direct investigations since his day will stand comparison with that in any branch of science. In this field the isolated observer has little place, less even than in meteorology, and it required the economic stimulus of submarine cable projects, abetted in our own day by that of the fishing industry to produce the government financial support without which the data cannot be obtained. The famous voyage of H.M.S. *Challenger* (1872–6) laid the foundations of our present knowledge; among many later expeditions that of the German ship *Meteor* in the South Atlantic (1925–7) has been one of the most fruitful.

From the *Challenger* reports—fifty volumes—published by the British Government, were derived our first outline picture of the form of the ocean basins, the character of their bottom deposits, the physics and chemistry of sea-water and the dynamics of the ocean circulation. They also provided a tremendous bulk of data on "hydro-biology." It is important to notice that physical oceanography and hydro-biology, though involving one another in important respects, are distinct specialist sections of the subject, the latter clearly falling within the field of biogeography.

On the physical side the chief advance of later years has been the perfecting of sonic- or echo-sounding, which has revealed the expected complexity of sub-oceanic relief. Of high interest also is the recognition of a broad structure within the ocean, analogous to that of the atmosphere and involving water-masses, distinctive in physical and chemical properties as well as in fauna and flora. The chemistry and bio-chemistry of sea-water has also come under detailed review and much has been learned of the relations of calcium compounds in the ocean, together with the related topics of dissolved gas content and hydrogen-ion concentration. The expedition in 1947-48 of the Swedish ship *Albatross,* which made use of the best modern equipment, is already adding further to our knowledge of the oceans.

A valuable but formidable summary of the field of modern oceanography has recently been written by Sverdrup, Johnson and Fleming. It serves both to measure the extent of the recent advance in knowledge and to indicate how far these exceed the bounds of "geographical" oceanography. It is of course true that much of the more recent research may have ultimate economic repercussions and thus illumine the earth as "the home of man." But the same is obviously true of electronics, radio-research and a host of other subjects which cannot fitly find place in the geographical curriculum.

It is idle to pretend, for instance, that bottom deposits and submarine relief are directly important elements in human environment, and much of marine hydro-biology is equally marginal to the geographer's chief interests. It is obviously dangerous to draw any hard and fast line here and the future may bring the oceans and their life much more closely within man's "living space." Meanwhile, the geographer's contact with oceanography chiefly concerns the oceanic circulation and its influence upon climate. Further, there is beginning to emerge a true regional geography of the ocean in terms of "water provinces" and life-regions and this will obviously find place in the geographer's general survey of the surface of our planet, Yet the oceans, whether regarded as a link or a barrier, show a unity which conceals and dominates regional variations; variations in water character or surface life-content are not factors in navigation or world commerce. With the single exception of the fishing industry, dependent on the distribution and migration of food-fish, differences between one part of the ocean and another are subordinate features within the field of general or economic geography. Other aspects of the science of the sea which are evidently important in "human ecology" are those concerning the origin and nature of the tides and the formation and distribution of sea-ice.

Our discussion of oceanography leads us to a general question of obvious moment—how far is physical geography, as such, a subject in itself and worthy of cultivation? The specialist sciences of Land, Air and Ocean are in some measure interdependent, though they diverge ever more widely in their detailed branches. Geology, indeed, is sometimes *defined* as if it included the other two, but it is not in fact practised in those terms. Its link with parts of oceanography is close but it has no such affinity with modern meteorology. Though wind and rain are potent geological agents, the geologist is little concerned with the physics of their origin. Meteorology and oceanography are closely interlocked along part of their common frontier, but hydro-biology is outside the meteorologist's ken.

In the light of these facts there is evidently a sense in which a balanced and compendious picture of lithosphere, hydrosphere and atmosphere, taken together, has a scientific rôle and value, quite apart from the demand made for it by geographers. Aside altogether from its human aspect, our planet and the inter-relationships of its parts are a unitary subject for study which presents a certain satisfying intellectual completeness.

It was such a concept which lay behind the useful subject "Physiography," popular in a former day. Though now outmoded by the growth of specialisms, physiography was essentially the geographical synthesis at the physical level; sometimes it was extended to cover an outline of biogeography and anthropo-geography, but never very successfully. Yet it attempted a synoptic survey of the realm of natue and its interlocking parts. It is no doubt true that the growth of knowledge has made the composition of such a "physiography" more difficult now than in the 19th century. But the unitary point of view still has its value and it is still possible to conceive a physiography which is at once both more and less than the aggregate of geology, meteorology and oceanography. The geographer must never cease to maintain that the proportions and relations of things are as much facts as the things themselves Modern physiography subserves the needs of biologists as well as geographers. Whether we regard our planet as the home of life or as the home of man, the sum total of its intricate physical economy is a fit and necessary background to the study.

It is for these reasons that specialization in physical geography is to be regarded not only as permissible but also as necessary. It is necessary in order that the solution of the central problems of geography itself shall be continually assisted by contact with the developments proceeding in the physical and biological sciences. But it is valuable also in enabling geographers so trained to lend appropriate assistance in the borderline fields of hydrography, forestry, agriculture, etc., and the solving of general problems concerned with the planning and conservation of "resources" which press so heavily upon the modern world.

It is a fair and pertinent question to ask the trained geographer what he can "do" in the field of practical affairs. His answer will inevitably be a two-fold one. As a humanist he is entitled to claim that the study has given him a philosophy and a point of view at least as valuable as that of the historian. As a scientist he can claim a synoptic knowledge of "Land" in all its aspects and one particularly relevant to our time and problems. But in the field of practical affairs he will be of little actual service unless his knowledge of physical geography is considerable and not limited to elementary borrowings designed merely to sketch a general background to a documented human picture.

Biogeography

When the geographer is in "blueprint mood" devising ideal schemes for his subject, he is naturally inclined to recognize a sub-division called biogeography. This seems logical enough, but reflection shows that it begs a very large question which may be tersely expressed, following Fairgrieve, as asking whether the central theme of geography is to be "man versus the rest" or "organic versus inorganic." The facts of plant and animal distribution and the existence of veritable communities of both plants and animals are in themselves facts of geography in that they help to differentiate the earth's surface and are relevant factors in human environment. But there are also quite literally fields of plant and animal geography kindred to that of human or anthropo-geography, concerned with the relations of plants and animals with their environments, both physical and biological. Evidently in its widest sense, as distinct from physical geography, bio-geography must include human geography. In working practice it is taken as concerned simply with plant and animal geography.

It is perfectly clear for the reasons noted above that the geographer is vitally concerned with the floras and faunas of the earth. It is no less true that the world of plants is more important to him than the world of animals. The immediate dependence of the higher plants upon soil and climate makes them valuable as "indices of terrain" and ensures delicate adaptation to physical conditions. The larger land animals, in general, show no such direct reponse to soil and climate, though they are closely linked in respect of their ultimate food supply with "vegetation" or plant communities. In this sense the green plant stands as intermediary between the inorganic and organic worlds; the power of green plants to manufacture the fundamental foodstuffs of animate creation gives them a highly central rôle in the economy of our planet.

All this and much more the geographer must of course appreciate; it would be unnecessary to make so obvious a statement if the facts and relations in question formed, as they should do, part of general education. Students approaching geography from the "Arts side" are not absolved from knowledge of such by the feeble plea that they "know no botany."

It is wise to recognize, however, that plant and animal geography, properly so-called, and regarded as fields in their own right, are divisions of botany and zoology respectively and cannot be pursued without an adequate knowledge of these sciences. This does not mean that the geographer can rest content with unorganized *ad hoc* borrowings from biology. We have sought to show in the case of physical geography that this method is untenable; it is not less so here. Not all the facts which, on a strict interpretation, are biogeographical, are equally important to the geographer, and his difficult task of selection still remains. But to understand the geography of plant and animals, like that of land-forms, involves the genetic approach.

The facts of distribution are not static or given facts, but the resultant of complex histories. They have developed and are developing. A general knowledge of the scope and methods of biogeography is necessary to all geographers, and it is desirable that those with suitable equipment shall continue to share the specialist fields involved. Without the help of such workers, much of significance to General Geography must be lost. All that we can attempt here is a brief sketch of some of the principles involved.

The subject of geo-botany, as some have termed it, embraces two distinct though not unrelated branches. On the one hand there is plant ecology, the study of plant communities and their relations to their environment, both physical and "self-made." On the other is the great group of problems which constitute "floristic" plant geography, concerned with the distribution of families, genera, and species and the controls or mechanisms by which the existing distributions have been brought about. In this field we are concerned not with vegetation in the large but with individual plants or groups of plants, believed to be related by descent.

The distributional facts are extremely manifold and complex. Some families show a wide—almost world-wide distribution, others are discontinuous, occurring in one or more separate and isolated areas. The distribution of some families extends across major climatic boundaries, i.e., they embrace genera adapted to a wide range of climatic habitats. Others—notable examples are the Palms—are compactly restricted to one of the major climatic realms. Genera show equally diverse distributional patterns. *Vaccinium* ranges throughout the North Temperate lands but extends also in parts of the Tropics. *Eucalyptus* is confined to the Australasian region, while the Magnolia tribe has a distribution which recalls, but by no means exactly parallels, that of the so-called "China Type" of climate. Larger scale maps of smaller areas, such as Britain, reveal similar facts and prompt like questionings. Of the five British species of the genus *Primula*, *P. vulgaris* occurs throughout the British Isles, *P. veris* is excluded only from parts of the western coastal lands, *P. scoticus* is confined to the Orkneys and Northern Scotland, *P. farinosa* to Northern England and South-east Scotland and *P. elatior* to the East Anglian region.

Confronted with facts such as these, the geographer, recognizing that plants are sensitive to climate, is tempted to use them as simple climatic indices. This is too facile a view and in some cases plainly at issue with the facts. In a wider sense vegetation is an evident register of climate; forest, grassland and desert show a pattern manifestly accordant in a general way with atmospheric conditions. But to use the same argument floristically is to overlook the historical element in the problem. We must view such distributions from the dynamic standpoint. They register plant migrations recently achieved and no doubt still in progress.

We live in the aftermath of the Pleistocene Ice Age during and after which great variations of climate have affected our planet. This itself was little more than a sequel to the great geographical transformations wrought by the mid-Tertiary movements throughout the world. Mountain ranges have been raised, plains sunk beneath the seas and the whole terrestrial pattern re-shaped while the ancestors of the present flora were evolving and migrating. The problem of the distribution of plants is thus in part a geological problem, and though the fossil record of Tertiary plants is very incomplete, it cannot be ignored. Problems such as these are evidently no part of the proper field of geography, but unless the geographer is aware of them his perspective is gravely in error.

It is clear enough that in this field the botanist, working as a plant geographer, is pursuing an investigation analogous to that of the human geographer. He, like the human geographer, ignores "historical geography" at his peril. Sir Halford Mackinder was wont repeatedly to use a simple example in explaining the two-fold nature of geographical explanations. A man standing on a hill may be said to be there because the hill is sustaining him there—or alternatively—because he walked or climbed there. In the same way "a primrose by the river's brim" is there because it thrives in, or at least is tolerant of, its physical environment. It is there not less certainly because its seed was able to reach the site and its seed-parents and their ancestors were enabled to reach the region. The background question is how and why this migration was achieved.

All this still leaves it true that plant distributions are facts of geography and, in so far as they are temporarily stable, they make for a definite differentiation of the earth's surface. Yet no great geographical illumination can be obtained from this fact. If we could conceive of the whole of the known facts of floristic plant distribution plotted on one vast map of the world, the only generalities that would emerge are well known to the geographer already.

The great plant realms, Boreal, Palæotropical, Neotropical, South African, Australian and Antarctic would stand out clearly enough, but many details of the distributions would be neither geographically informative nor self-explanatory. Evidently some of the detailed facts are highly significant geographically, particularly the distribution of plants of economic value—e.g., rubber. But no one can pretend that the presence of *P. elatior* in East Anglia and its absence in the Midlands is a major element in characterizing either region. It *might* reflect the influence of climate or soil conditions, but there are obvious alternative ways of establishing the contrasts between the two regions in these respects.

The other aspect of geo-botany, plant ecology, opens up a group of enquiries closely analogous to those pursued by the geographer himself and, significantly enough, exposed to somewhat similar doubts and criticisms. We may quote the question imputed by A. G. Tansley to those botanists who doubt the status or validity of plant ecology. "Plants yes, but vegetation, what is it but aggregates of plants?" Here the geographer is not in need of conviction and can cordially assent to Tansley's reply. "The state of mind that is expressed in such a question shows a failure to recognize that because plants not only live in aggregates but have been developed and moulded by the gregarious life, their behaviour cannot be understood unless these aggregates are studied for their own sakes." To this the geographer might add that, from his own point of view, the aggregate of plants occupies an area of land which also requires study for its own sake.

Plant ecology rests on the fact that vegetation in the mass

consists of plant communities, recognizable and in general mappable units, sharing a common habitat. Ecological study must include the quantitative description and analysis of the community and a close review of the habitat factors. These are respectively climatic, physiographic (slope, aspect, etc.), edaphic (soil) and biotic. The last includes the influence of the associated animals, including man, and also, in the wider sense, the mutual influence of the plants on each other. These last, however, really constitute the self-made or internal environment as distinct from the external or geographical environment.

A plant community has an inherent life or tendency to development of its own. The American botanist, Clements, transformed the outlook of plant ecology when he established his conception of a "climatic climax." The climax community is "dominated by the largest and particularly the tallest plants which can flourish under the climatic conditions prevailing." The conception is most readily grasped by considering the succession of communities which follow one another in the stages of re-colonization of a piece of bare land. The pioneers, generally lowly plants, by their activity in life and decay after death, literally pave the way for successors and so by stages the area becomes occupied by an ever richer or more varied flora culminating in the most fully developed example of such which the climate permits—an essentially stable condition.

Here once more is the dynamic or kinetic point of view in regard to vegetation; without it the study of vegetation, whether by botanist or geographer, is liable to crude misconception. By simple observation one learns that wide areas of sandy or gravelly soils in England give rise to heath country; elsewhere birch woods come in, without any obvious change in the habitat conditions. In fact the climax community in this case is believed to be dry oak wood, i.e., oak wood dominated by *Quercus sessiliflora*. The existing communities mark various stages in the succession towards this climax. Heather and its associates come first and prepare the way for the birch, which finally gives place to the oak. Mixed and intermediate stages inevitably abound.

In closely settled and cultivated country the climatic climax is a theoretical concept since it is nowhere attained in unmodified form. Thus the climatic climax over the greater part of the English Lowland would be "oak-beech summer deciduous forest" of which certainly no unmodified remnant remains. Geographers, who necessarily deal with the country as it is and not as it was, therefore welcome Tansley's extension of the term climax to cover the case of any relatively stable community which may be regarded as the end-term of a succession. The full climax may be inhibited by a variety of factors including human interference, but it is still useful to think of it in a suitably qualified local and temporary sense as a climax.

As our knowledge of plant communities grows, it is increasingly clear that human impress on the vegetation cover has been so incessant and widespread that natural vegetation in the ideal sense is to be found with difficulty. In settled lands all semblance of the original plant cover is gone. The thickest Sussex oak wood is no "virgin forest" but woodland of comparatively recent growth. The famous Broadbalk Wilderness at Rothamsted is a standing witness to the vigour and speed of the regeneration of woodlands. Cultivated land before the harvest of 1882, it was then abandoned and in less than seventy years it has become a closely grown thicket of hawthorn, oak, ash, and sycamore.

In conclusion, it may be pertinent to remark that not the least of the geographer's debts to biogeography lies in the evident arguments from analogy and the question, which must ever stimulate the mature student of the subject, as to how far they are valid. Here, indeed, we resume for a moment our earlier discussion in Chapter II. In some degree, wisely or not, the geographer is wont to claim the rôle of human ecologist. It is, of course, abundantly clear that he does not claim the whole of that field. The "internal" relations of human communities, as recorded by history and analysed by the social sciences, are not primarily his concern, nor will anyone admit his competence to cope with them. It is only when he limits the problem to studying the influence of Man on Land, not of Land on Man, that it is plainly necessary for him to be informed as to the "internal" economy and development of human communities.

A developing human community, no less than a plant community, transforms its habitat. It changes the face and modifies the pattern of "places" or areas. Here is the geographer's warrant for keeping his place in the historico-social team for the sake of what he gets, if not for what he gives. But though what we may term the "positive problem" of geography, the influence of Land on Man, is full of dangers for the unwary and though its study has been confused and discredited by hasty and uncritical work in the past, it remains a problem to which geography may hope to return. But at its very heart lies the problem "What is Man?" When we note the regularities and recurrences of settlement patterns and kindred phenomena, we may feel at times that we are dealing with phenomena not fundamentally different in type from those of the plant and animal ecologists.

CHAPTER IV

GEOGRAPHY AND MAPS

And what is there in all the known world which maps and authors cannot instruct a man in as perfectly as his own eyes?

BISHOP HALL (1605).

"IN geography," wrote Dr. H. R. Mill, "we may take it as an axiom that what cannot be mapped cannot be described." This evidently follows, if we regard geography simply and sanely as the description of the surface of the earth and its "areal differentiation." The map becomes pre-eminently the geographer's tool both in investigation of his problems and the presentation of his results. This is evidently not to say that he alone is professionally concerned with maps; they are part of the equipment of civilized life and are used of necessity in many branches of learning. Still less does it mean that we can confound geography in the large with cartography. The geographer must be adequately trained in cartography, as in many other crafts, but the making of maps in the cartographic sense is not his main concern.

The geographer's prime need of maps is twofold. In the first place it is impossible for him personally to visit and inspect the whole surface of the earth. Travel is indeed vital to his work, but the travel of a lifetime would not replace the pictures given by maps. From this there sometimes arises a criticism that the geographer's work is "second-hand." In a sense the ground is the primary document, the map a secondary document, necessarily partial and imperfect. The criticism has a certain weight if our geography consists almost wholly of map-knowledge of distant areas. Surely, one might think, no study by professional geographers of maps of British Columbia can produce knowledge equal in validity, reality and detail with that of an observer living in and travelling within that province.

In some measure this is undeniable, but it is by no means wholly true. If the man on the spot were a geographer, it might indeed be true, although even then he would need maps to crystallize and codify his knowledge. But the cardinal fact so easily lost sight of is that the map correctly used can display patterns and bring to light relations, perceived with difficulty or not at all by one living on the spot. The point need not be laboured, for it is so readily demonstrated in a simple instance. One might live for a long time in a close-textured, complex and heavily wooded landscape without gaining a complete view of its pattern such as might be got in a leisurely flight across it. Aerial photography affords within certain special limitations the most compendious picture of "ground" and the most rapid method of surveying it. Or again we may recall how inadequate our eyes and our legs prove as surveying instruments in a strange town. A quick-witted mobile urchin may in one sense "know his geography" as he conducts us by tortuous by-ways from station to hotel, but neither he nor we have any adequate picture of the pattern of the town, without the benefit of maps.

These are elementary points, but for our present purpose they need considerable emphasis. In the basic cartographical sense the geographer does not make maps, leaving this task to the land surveyor or the civil engineer. He is, however, a past-master at using maps, distilling the ultimate significance from them and re-shaping or re-arranging them to display relationships.

The topographical maps of the British Ordnance Survey have no superiors in the world. They achieve an objective picture, subject to their adopted conventions, of the geography of our country. But it is in no sense necessary that the men, who survey the ground, draw the map, engrave, photograph, and print the plates, shall be in the academic sense geographers. The geographical task, the interpretation of the map, remains to be done when this work is finished. It is, primarily, to describe and interpret the ground of which the map is a partial picture.

Here we must evidently make a distinction between the work of the geographer in well-mapped and ill-mapped regions. For the greater part of the land surface of the globe there are no accurate maps on scales of the order 1 : 50,000, comparable, say, with the British "One Inch" map. Nothing less than these will suffice as the basis for the geographer's full synthesis. Where he is denied them he must perforce use the best he can get and he may then require to supplement them considerably by his own efforts, using surveying methods to locate and portray features he wishes to discuss. Just so may the working chemist become on occasion his own carpenter and glass-blower, but carpentry and glass-blowing are not chemistry.

It is then quite essential that the student of geography should clearly perceive the bearing of the two main divisions of what is commonly called "map-work." On the one hand, there is the study of the cartographic properties or characteristics of maps, and on the other, the more difficult and important task of interpreting the map as a picture of the ground and of supplementing and adapting it by further study.

It is essential that the geographer should know the quality and limitations of his tools, in the same sense as, say, a musician or a golf professional does. One road to such knowledge is a study of how maps are made. The first step in logical sequence is Survey. We cannot here discuss surveying methods at length; they are fully dealt with in numerous accessible works.

It is worth remarking, however, that, in a very real sense, surveying is an art, not a science. Its principles are few and simple, its practice infinitely skilful and remarkably accurate. In a very few minutes it can be demonstrated how by measuring a base and building upon it a system of triangles, points can be rigidly fixed—the apices of these triangles. Now these triangles can be broken down into smaller triangles, till at length a large number of fixed points are available, by reference to which the topographical features can be inserted by simple measurement of either angles or distances. The process is exactly the same whether we use a chain, a compass, a plane-table or a theodolite. The introduction of the clinometer for measuring slopes, or the level, or less accurately, the aneroid barometer for determining height, complete the instrumental picture, at least to the point at which we can see that the theodolite with its horizontal and vertical circles is the sum and summit of the surveyor's equipment, complete in itself and capable of delicate adjustment.

The student of geography practises at least the simpler operations of Survey for two reasons. In the first place they assist the understanding of full topographical survey. In carrying out the survey of a small area it is possible to use essentially the same plan as would be appropriate to a major operation. There must be preliminary reconnaissance, the fixing and marking of the selected triangulation points, the measurement of one or more bases, the carrying out of the primary triangulation (whether by compass, plane-table or theodolite), the determination of the heights of selected points, and finally, actual mapping by plane-table, working on a basis of the determined triangulation points.

Full and formal survey would involve, it is true, certain other operations, the determination of "mean sea-level" to which heights are referred, the carrying out of astronomical "fixes" (latitude, longitude and azimuth at certain points) so as to place the survey on the surface of the earth as a whole. For large areas various reductions and corrections will have to be applied. The base will have to be "reduced to sea-level," and allowance made for the curvature of the earth's surface, in calculating the triangulation. But in principle at least, if not in detailed practice, our small-scale survey is a full and fair microcosm of the larger operation and can teach us much about it. The knowledge is essential in assessing the "rank" or validity of published maps.

We are dissatisfied with our own surveying efforts because of their palpable crudity and inaccuracy. We soon learn that in mapping, the part cannot really subsist without the whole and that "floating maps" with few or no astronomical fixes and not rigidly set in a major triangulation are of little value or accuracy. It is then with something of a shock that we realize that much of the world is still unmapped in any proper sense. Its outlines are based on the compilation and reduction of reconnaissance surveys which do not really fit. In illustration we may cite A. R. Hinks who wrote, only twenty-five years ago: "When the Union of South Africa wanted to make a small relief map of Natal to be shown at Wembley last year, they sent round to the Geographical Society for material, and were staggered to find that we could not supply it, for the good reason that Natal has never been mapped." He added that geographers "were getting tired of asking whether there is yet a single topographical sheet to be bought in Australia. I believe that the answer is still No!"[1]

Of Canada he wrote: "Thanks to the labours of Mr. Wallace we do know at last, what Lord Southesk never did, where he went on his journey of 1859 in the Rockies, but there is still no published map good enough to show the route upon." To many these will be surprising facts and they could be multiplied almost without limit. We are led to recognize that though such regions figure in our atlas and wall maps, they figure as little more than sketch-maps though our attention is rarely drawn to the fact. And do not let us claim that "they are surely accurate enough for practical purposes," for if the practical purposes are those of the geographer the answer is No!

The further value of elementary survey to the geographer, as to other field observers, is in enabling him to add data to an existing map. In a well-mapped country such as our own there would seem at first sight little need for this, but the seeming completeness of the map is illusory. It shows only a selection of the visible features and in any case can show only what is known or seen to be there at the time of the survey. Plant communities, archæological sites and many other features may need to be located and portrayed and, although this is a simple matter in closely settled or cultivated ground with its innumerable buildings, woods and field boundaries, our open moorlands above 1,000 ft. are almost featureless on the published maps and even their contours are widely spaced.

A recent exercise carried out by a party of students may serve as illustration. An area of high interest to the physical geographer is Water End in Hertfordshire, where the head-

[1] Needless to say this state of affairs no longer exists, for the more developed parts of the Commonwealth have now topographical maps of various scales.

waters of the river Colne pass underground in a series of swallow-holes, past and present, by which the surface waters have descended. To map such features, maps on a scale of 1 : 2,500 (25 in. to the mile) or larger are required. With the aid of the existing 1 : 2,500 map it is a simple matter by means of plane-tabling to show the distribution of the swallow-holes and the form of the major depression in which they occur. Similarly, accurate work upon stream profiles starts from existing maps, but involves adding details to them.

Another great branch of what may be called mathematical cartography is that concerned with map projections—the problem of balancing the inevitable errors involved in portraying the curved surface of the earth on flat sheets of paper. From this standpoint it readily appears that there is no such thing as a completely accurate map, for the problem involved is incapable of ideal solution.

The difficulty concerns maps of all scales, but is chiefly conspicuous on maps of smaller scale.

Relative to the earth as a whole, position is fixed by the so-called geographical co-ordinates, latitude and longitude. The problem thus becomes that of representing the parallels and meridians of the globe upon a flat sheet of paper and there is literally an infinite number of ways of doing this. Mathematically speaking, we are dealing with only a limited series of special cases of a general problem, that of representing the "pattern" of a surface of any given form upon some other surface and in the general case the mathematical complexities are considerable. But our cartographic aim is simple and definitive. There are only a couple of dozen map projections which are significant and the number which in the narrow sense are useful is smaller still. The needs of the geographical student in this matter are adequately summarized by Hinks as follows: "It is important to have a clear geographical or numerical idea of the merits and defects of each (projection), to be able to decide at once on its suitability for a given map, or when one finds it actually employed on a map to recognize what a map so constructed will do and what it will not do."

As to the place that should be taken by a knowledge of Surveying and Map Projections in the training of the modern geographer, opinions have differed widely. On the one hand, since geography is attractive to students of the humanities, who bring with them often the common fear of and aversion from mathematics, it is often contended that these subjects are "not geography" and that a slight or perfunctory study of them is adequate. In its extreme form this view is open to grave criticism and does the status of the subject much harm.

Evidently, on the contrary, these branches of cartography are geography at least in the literal sense. No one claiming the title of geographer, however humbly, is entitled to be ignorant of how maps are made. Nor can it seem other than incongruous if the geographer cannot make at least a rough map, or cannot determine his latitude, or if he sins against the elementary light of "map projections" by measuring directions on a Bonne and areas on a Mercator projection. He needs also to be very careful in measuring distances on maps of small scale on any projection whatsoever.

To concede this is not, however, to admit that these matters are central in geographical scholarship; they are rather ancillary. It is an extreme and indefensible view to regard map projections as "the most important part of geography" and it implies a lamentable ignorance of what geography really is. The purist, insisting on the root meaning of the word, might seem to have a case, but he must none the less be regarded as a pedant. The question may perhaps be best summed up by saying that it is not in his knowledge or treatment of Survey and Projections that the quality and scholarship of a modern geographer are chiefly seen. We need place no rigid limits to what is the degree of desirable acquaintance with such topics geographers should attain. But we must recognize that these branches of cartography constitute a complete and self-subsisting field of their own with its own point of view and its own criteria of scholarship. The final and refined objective of survey is the measurement of the earth—the minutely accurate determination of its size and shape as studied by geodesy.

A National Survey is not only the basis of necessary and practically useful maps. It makes, if conducted with sufficient refinement, its contribution to the figure of the earth, the computed sea-level surface or "geoid" which is the ultimate basis of all "mapping." Geodesy, like any constituted subject, has its own internal disputations and issues of which the acuteness and gravity are hardly to be appreciated by other specialists, and certainly not by geographers. Thus Hinks, in an address which we have already quoted, after praising the pioneer geodetic work of Col. A. R. Clarke and noting the lack of recognition accorded to it, wrote: "I shall regret as long as I live that the British Delegation at Madrid last Autumn did not make a concerted fight for Clarke's figure of 1880 as the Standard Figure of the Earth. . . . But a grave mistake was made that day when it was decided by the narrowest of majorities that a brand new figure of the earth, derived from a restricted part of it, by a process which is the subject of controversy, should be imposed on a generally reluctant world as the collective wisdom of a plenary International Conference."

Here evidently are weighty matters on which judgments may differ and feelings run high. But the geographer cannot conceal from himself that the matters at issue barely touch him at all. The difference between rival figures of the earth is quite immaterial in his habitual field of work.

We are far from exhausting the subject of cartographic characteristics when we have acquired an elementary knowledge of Survey and Projections; indeed, in one sense we have barely approached the matter. The map, properly so-called, is still to be made when the long preparatory stages of survey, triangulation and computation have finished and all questions of choice of map projection are settled. In its finished form it must needs embody conventional signs for physical or man-made features, such as are summarized in the "characteristic sheet" of a series of maps, or in selected and condensed form in its marginal legend. The elementary stages of map-reading involve familiarity with these characteristics. The matter is at once important and in a sense trivial. There is here none of the technical difficulty of "mathematical" cartography, but there yet remains a true art in map-making—the art of clear and effective portrayal of distributed, juxtaposed or even overlapping features. It cannot be studied or discussed in the abstract but only by reference to actual instances.

All this, however, is the mere grammar of map work and, though a necessary preliminary, can be made too much of. Younger students and many "laymen" are under the impression they can read a map when all they are doing is labouring to spell it out. Just as the practised reader of print can take in whole words, sentences, or even paragraphs at a glance, so the real map-reader is master of the geographer's essential art of seeing things together. The best road to conviction here is to do a little map analysis, as it is commonly called, i.e., trace off and study separately the various categories of features which the map presents: contours, streams, routes, buildings, etc. This is laborious work and at first seems barely worth the trouble, but only so does it become clear that casual scrutiny of the map conveys a view not of the whole but the parts in succession.

The average person when claiming ability to read a map generally implies no more than his power to take, as it were, the second on the right and the third on the left and thus reach his desired objective. Such "route-finding" involves little more than the childish rudiments of map-reading, and only the cartographically illiterate mistake parish boundaries for footpaths or try to cross rivers where no bridge exists.

Map-analysis tracings immediately bring to light features and patterns, which would escape the inexpert map-reader, even after long scrutiny and a lifetime of "finding his way about" on the map. The "contour and water" of the map give a picture surprisingly graphic and clear and prompt the reflection that, in the narrow sense, this is all a "topographical map" is designed to show. The attempt to add the additional data of human "occupancy" at best confuses and at worst hides almost completely the physical picture. Even more interesting, because less familiar, are some of the human distributions. A "house" or "building" map reveals quite definite but unsuspected patterns bringing out distinct areas in which the settlement pattern is "nucleated" with clustered buildings, and others with "dispersed" settlements—single buildings or small groups. Such a map, in a rural area, gives the best possible picture of certain aspects of the distribution of population, even though a small and unknown proportion of the buildings shown are not habitations.

Most significant of all, in many cases, is the route net studied separately. Here it is important, in the first instance, to treat all routes as of equal status and represent them by single lines. One of the chief faults of many series of topographic maps is their over-emphasis of main and other "graded" roads. The practical reasons for such treatment are clear enough. The public as "route-finders" value just such emphasis. But the real significance of the pattern is often lost thereby. Old routes may be marked today only by field-paths or unmetalled lanes, and surprising alignments spring to the eye when the single-line method of tracing is adopted. Field-boundaries themselves have proved of like significance in the discovery of Roman roads, but these are details of pattern requiring larger scales for study.

All such forms of map-analysis may help in the special investigation of a piece of country but, in training, their chief use is in demonstrating to the eye what it does not, at first, readily see on the full map, and in assisting the learner to develop the art of true map-reading. There is clearly no other virtue in taking apart the several elements of an effective and well-compounded map. Its whole purpose is to bring the elements together so that they can be seen in relationship not in superimposed tracings but on the single sheet. Part of the difficulty of seeing them in relationship is artificial, residing in the limitation of the map itself. Each successive layer of information must necessarily obscure part of its predecessors. This part of the difficulty can be attacked only by experimenting with other and better conventions and methods of portrayal, but it is inherent and cannot be entirely overcome. The more important part of the difficulty is in the reader's lack of this skill which can be acquired only by hard study and practice: it is here that map-analysis is so important an aid.

Teachers and examiners of geography know well how stringent a test of general and geographical ability is afforded by the simple-sounding question: "Write a description of the country shown on the accompanying map." At the simplest level this is quite literally an exercise in translation from the "foreign" cartographical language into English. It is when the account is intended to include not translation only but interpretation that the task gains interest and difficulty.

Here there are evidently two variables, the quality of the map and the knowledge of the reader. The most expert geographic eye cannot obtain more from a crude map of a quite "unknown" terrain than has been put into the map. With more detailed and accurate maps the interpretation obviously gains in complexity, but if no background knowledge of similar regions can be employed, the results are still limited to translation. At the best, such interpretation can be little more than a sort of detective game and too much emphasis has sometimes been placed on this aspect in elementary training.

What the reader often requires to know is, "Am I to deduce my conclusions from the evidence *on the map* or to supplement the evidence by other knowledge?" It is no doubt a useful exercise on occasion to confine oneself to the mapped information and particularly useful as a preparation for field work in the area. But map-interpretation at its highest level involves a general knowledge of the terrain. Suppose that one had spent a period in field study of an area, without the help of good large-scale maps; one would obtain a first-hand knowledge of the appearance and character of hills and valleys, forests, fields, settlements and routes. If then the maps were produced and studied, two lessons would immediately be learnt. On the one hand, the map is an imperfect picture of the ground, leaving much to the imagination, and giving sometimes no more than faint hints of the realities of the landscape. On the other, it at once brings to light patterns and relationships overlooked in field study.

This important point, familiar to all geographers, is clearly stated by Preston James: "In geography the phenomena to be described and understood are much larger than the observer—it is actually the details that are obvious: but the broader pattern and relationships can only be studied or described by reducing them to observable size on maps. To a very small creature living on the surface of a half-tone photograph the detail of the printed dots would become quite familiar . . . yet the larger pattern of those dots which are combined in the general areas of light and shade to form the lineaments of a picture would not really be at all obvious."

These reflections enable us to perceive something of the rôle which map interpretation should play in the study of regional geography and in field work. It might seem that the advanced study of regional geography could not dispense with first-hand study of the regions themselves. As a desideratum this is practically unattainable on any full scale. It would, no doubt, be an admirable addition to geographical training if students enjoyed world cruises or, better still, studied their world extensively by air travel. Still better, evidently, would be a term's residence in any region prescribed for detailed study. But it is important to recognize that *large-scale* maps are not only the best substitute for study of the ground but that they have their own part to play in contributing to the understanding of a region.

It is a salutary and indeed an indispensable exercise to study in succession the "version" of a region such as the North Italian Plain on an ordinary atlas map (on a scale, say, of 1 : 3 million), on the International Map (1 : 1 million) and on a block of sheets of the Austrian Staff map (1 : 200,000). It is on maps of about the scale of the last, comparable with that of the British quarter-inch maps, that the detailed realities of the landscape begin to appear. Relief can be shown without drastic generalization, some aspects of land utilization can appear and, above all, the real texture of settlement comes effectively into the picture. At the same time, the effect of *general* pattern is not lost. Such study can be carried further for select or typical areas on maps of about the scale of 1 : 50,000. Here something like full physiographic accuracy is attainable, demographic detail is increased, and for closely settled or urbanized areas this is almost the minimum effective scale for study.

Such study of regions with the help of topographic maps on scales of about 1 : 200,000 or larger is an essential part of the method of regional geography. A few hours study of such maps conveys more real knowledge of a region than the same time spent in reading or listening to lectures. But the interpretation will be limited unless we bring to it a knowledge of the physical and human history of the area and the economy of the people who inhabit it. All this is the pre-requisite background knowledge needed to enable us to obtain from the maps their full significance. Further, our past study of the maps will certainly suggest further problems for the solution of which we shall go back to the geological, historical and economic literature of the area.

The map, in short, answers some questions and asks others. It forms our starting point in regional study, summing up what is accurately known and accepted, but directing our further enquiries. When these are complete we can, mentally at least, re-draw the map, filling in the void and elaborating its detail. This and nothing less is the true art of map-interpretation.

Similarly there is a right and wrong way to use the map in the field. As we have noted, the ground itself is the primary document, and in the field we should work from the ground to the map. A simple illustration of the principle involved is given in the training of air navigators. The wrong method of fixing one's position from the air is to study the map first and then to seek to "make the ground patterns fit." The airman off his course is better advised to study the ground until he picks up a point—road and railway crossing, stream meander, etc.—so distinctive that he can then identify it on the map without doubt. This is no more than a simple if effective practical "tip." More generally, however, in studying country the same method must apply.

Geographical field work involves many and various enquiries which we cannot here discuss, but its beginning is the comparison of the map with the ground. The map is a summary of certain features of the ground, but if our eye is too continually on the map, we tend to rest content with those features which the map shows and ignore the rest. From any hill-top vantage point it is better first to study the ground carefully and at length before returning to the map with its partial and selective presentation. It is, of course, true that some of the conspicuous omissions of the topographical sheet are supplied by the corresponding geological and other maps (e.g., land utilization maps). It is an excellent laboratory exercise to compare these maps, but when in the field the ground "integrates" all of them so far as visible and mappable features are concerned; it shows, too, further features which are not represented on any of them.

We come lastly to the important question of the special maps used by geographers in the investigation or demonstration of particular points or aspects. It is essential to realize that many of these are not, in the strict sense, maps at all, but cartographic diagrams or "cartograms." A map, properly speaking, is a picture, to scale, of the ground and its visible features. Many other phenomena can be mapped, with results highly suggestive or misleading according to the methods used.

A first and radical distinction is between phenomena which show continuous and discontinuous variation respectively. To the former the isopleth method is applicable as in the familiar instance of contours, isobars, isotherms, etc. The isopleth method is accurate and definitive in the measure of the amount and accuracy of the data from which it is constructed. But here we encounter the question of "critical values" and the problem of selection of data. A map designed to show that New York is situated beyond the zone of extreme summer heat in North America might show the isotherm 75° F. beyond the northernmost point of which New York falls. But why 75° F.? If we had taken the more commonly used 70° F. line, New York would fall south of it; the critical temperature in the view of most climatologists is 68° F., for which the isotherm is even farther north.

A signal and important example of discontinuous distribution is that of population. Generally speaking, a very effective portrayal of population or of various forms of production is afforded by the now familiar "unit-dot" method. If both the unit and the "areas of reference" (parish, county, etc.) are sufficiently small, unit dots can be distributed evenly over the areas, excluding only any quite uninhabited parts. In this way, by reduction of the scale of the final map, the effect of varying density can be effectively shown.

Alternatively, both population and crop or other figures

can be plotted with reference to the *known* general distribution, that is, the dots can be placed where we know or think they ought to go. Rashly used, this method "begs the question," but within proper limits it is valid. If a house or building map is used as a basis, by assigning so many persons to each house, the method is perhaps at its best, though "residual" dots will still have to be accommodated.

Despite its seeming precision, "dot mapping" can be rendered misleading or even ridiculous by an ill-judged relationship of "unit" to scale. Thus if we wished to show an estimated population of a country largely nomadic in a semi-arid area, we should not be justified in planting two dots representing 500,000 people each, within an area as large, say, as Arabia. In any case, the dot method breaks down with urban populations. Any unit small enough to show a rural distribution is many times too small to avail for the towns within it. Here indeed is the major discontinuity in population distribution and nothing can avail but the concurrent use of two different methods. The practice is growing of representing town populations by spheres drawn in perspective, but with radii which give a spherical volume proportional to the population.

Density distributions, whether of population or production involving arbitrary administrative areas, present further obvious problems, since the actual changes of density rarely coincide with the administrative boundaries. The resulting "patchwork" quality of such maps can be avoided in part by reducing to a minimum the number of density grades shown. Thus we may choose to distinguish only those rural units which exceed a selected density. This is permissible, under safeguard, to establish a particular point or to demonstrate a single relationship, but it is a dubious expedient and liable to abuse. Here again it is a question of critical or limiting values, and by biased choice of these the map may be made to lie as readily as the statistics on which it is based.

Isopleth, dot and "density" maps, despite their power to mislead if maladroitly used, necessarily retain something of a definitive and quantitative character. Cruder cartograms involving pictographic devices are increasingly common in the modern world. We may, for instance, show the distribution of public telephones within a county by printing a drawing of a telephone instrument as a convention. This is only mildly silly and over-elaborate, but similar devices have been pressed into the service of propaganda. Many will recall the cartograms purporting to show the "encirclement" of Germany, comprising a ring of guns pointing towards her frontiers, each of a size proportional to the population of the neighbouring State.

All these we may dismiss with disapproval. They prompt the reflection that the basic rule for the geographer in seeking a criterion to "control" his cartogram is a simple one. Since into the construction of any such device a subjective element must enter, it is vital to keep the cartogram as nearly as possible an actual map; its lines or boundaries coinciding closely with visible or verifiable features of the actual ground. As soon as this principle of guidance is abandoned the door is open to distortion and misrepresentation. The geographer is, of course, entitled to assume a minimum of common sense among those of his own craft and of the public who use his cartograms, but he must not rate this minimum too high. Neither of maps properly so-called nor, certainly, of cartograms is Bishop Hall's statement, with which we head this chapter, at all completely true.

CHAPTER V

HISTORICAL GEOGRAPHY

It (historical geography) is a study of the *historical present*; . . . the geographer has to try and put himself back into the present that existed, let us say, one thousand or two thousand years ago; he has got to try and think of the geography of that time complete; he has to try and restore it.

SIR H. J. MACKINDER.

ALTHOUGH the studies of geography and history are generally recognized as cognate and indeed, as Dr. Heylyn put it, are threatened, if parted, with certain shipwreck, British educational practice has always separated them sharply. In writing, as in teaching, each study has been usually conceived as a separate field, and their common borderland remains relatively unploughed. Yet both the geographer and the historian are well aware that these two studies are inter-related, and that each can, and in certain problems must, seek illumination from the other.

On the one hand, the historian, in his attempts to explain the location of past events, contrasts in agrarian systems, the migrations of peoples, the origin and growth of cities, military and naval strategy and the means of communication and transport from place to place, encounters problems to the solution of which knowledge of the geographical background is indispensable. On the other hand, the geographer, concerned as he is primarily with the transient present, finds himself continually faced with questions to which history holds the solution.

It is not surprising, therefore, that in our universities several specialist studies have sprung up to investigate the inter-relations of geography and history. Two of these investigations, which can be clearly distinguished in theory, though in practice their distinctness tends to blur, are relevant to our discussion. The one, which was the first to find exponents in England and is associated with the names of such pioneers as J. R. Green and E. A. Freeman at Oxford, is concerned with the geographical factor in history. The purpose of this kind of enquiry into what may rightly be called "the geography of history" is to advance the study of history by discovering in what ways and with what effect geographical conditions, various and diverse as they were from one area and from one time to another, affected historical phenomena. The other, in contrast, seeks to clarify present-day geography by resort to the materials of human history or, where these are available, to the findings of archæologists and historians.

It is this latter study, which we regard as an essential auxiliary to the understanding of contemporary geography, that we shall here discuss. To distinguish it from "the geography of history" its practitioners call it "historical geography," and it has for its main task the reconstruction of past geographies. We shall try to show that an adequate understanding of the geography of any area today demands at every turn enquiry into the processes which created that geography. We have already argued that knowledge of past physical events is necessary to understand the present physical geography of areas; similarly, the processes recorded in human history, which have fashioned their human geography, must also be understood. And the study of these processes, for the light which they shed on the world about us, is the ultimate, if not the immediate, purpose of the historical geographer.

"The geography of Britain," wrote Sir H. J. Mackinder in a classic work, "is in fact the intricate product of a continuous history, geological and human." And although it might be objected that the important factor of climate is not wholly subsumed within geological history, this generalization underlines forcefully the essential unity of geography as the outcome of processes operative throughout time. If the main objective of geography is the recognition and study of regions, as these are distinguishable in terms of physical and human criteria, then in furtherance of this end account must be taken of both their physical and human history. Whatsoever the region, unless it has wholly escaped human impress, it is a compound of two contrasting yet interwoven elements—a physical basis, which is relatively stable, and cultural patterns and distributions which are always changing.

The present-day geography of any settled area embodies residuary features of many past geographies as it inherits the momentum of forces generated in the past. An area of land, like an individual, is not fully explicable except in terms of its history. When we try to understand by means of analysis the composite features etched by nature and by man on the face of our countryside, we need to know when and how they first appeared.

Geography is in fact inseparable from the history which produced it. The intelligent geographer studies not merely country which is conceived as momentarily static, but also the history of country which is still being enacted before his eyes. Curiously enough, and evidence of the relative belatedness of the historical geographer's viewpoint, we often search in vain in the many available histories of countries for an account of how the face of the countryside changed, although notable exceptions leap to the mind such as A. E. Zimmern's *The Greek Commonwealth*, G. Hanotaux's *La France en* 1614 and Sir J. H. Clapham's *The Railway Age*. This usual deficiency can be made good by the study of historical geography. And, like history itself as many conceive it, this is no mere antiquarian pursuit, a search for past forgotten things which might well be left in obscurity; rather it is concerned with present questions for which the past holds the clues. If we may take out their context words which aptly express this idea, we may say that historical geography attempts "to summon the living, not to invoke a corpse, and to see from a new angle the problems of our own age, by widening the experience brought to their consideration."

A great historian, F. A. Maitland, a generation ago, described the topographic maps of England as "palimpsests" or documents of our rural history, and was able to show that by deciphering them light could be shed on some of the problems of Domesday Book (A.D. 1086). This conception of the countryside or the large-scale map as a document is no less valuable and stimulating to the enquiring geographer.

Our topographic maps, which depict the countryside as it is today, or to be more accurate, as it was at the time of their production, are, like geological maps, made up of many period pieces. To the untrained eye a large-scale geological map presents a chaos of colour; only when the time sequence of the different rock formations there represented is known, can the student begin to understand the geomorphology of the area mapped. In the same way, the topographical map, which expresses visually the age-long impress of man and his works, may appear at first sight baffling in its complexity. But the student who has learnt through which distinctive phases of human history, and with what effects, a particular area has passed, is able to resolve order out of apparent chaos and to distinguish significant patterns and distributions in what seems to the untrained eye a confused and disordered scene.

Cities, earthworks, ridgeways, railways, farms, field patterns, mines and quarries—these and other man-made features indicated on the large-scale maps of the Ordnance Survey—reveal today definable relationships with their physical settings which it is the geographer's business to establish. We would insist that this is only part of his task, which would remain uncompleted unless he continued to examine these features in the light of their origin. We should think ill of the archæologist who was content to describe the time sequence and cultural characteristics of excavated finds and ignored their locational and distributional aspects.

Similarly the geographer's task is many-sided. He should envisage the present-day countryside of a long settled country such as Britain as a dynamic, not a static, entity. It may be convenient to regard it as temporarily at rest, as it appears on large-scale maps, yet despite the relative stability of its physique it bears a continually and inevitably changing expression which registers the effects of its continuous adaptation by man. "The map," wrote Mikhaylov in his *Soviet Geography*, "is a social document. It fixes time with its symbolism, alternation of colours, and peculiarity of design. The lines on the map are the handwriting of History."

As writers of geographical text-books know very well, the present becomes very soon the past. An up-to-date knowledge of geographical facts has its obvious value in relation to present problems; so also have those forces and decisions of the past which still affect our environment. The complete geographer, we would affirm, must be equally cognizant of the architecture of nature and of the artifacts of man. He should try to conceive the particular area which he is studying, whether by maps or in the field, in four-dimensional terms. Time, no less than the other three dimensions, characterizes the personality of a region. Except for such lands as the Polar regions or parts of equatorial Brazil where men have not yet made permanent homes, the geographer is concerned with cultural landscapes which have attained varying stages of development.

W. M. Davis a generation ago regarded landscape as a function of structure, process and stage, and although this definition related to the physical landscape, it is equally applicable to the human. A common interest in time scales (of very unequal length) unites the geomorphologist and the historical geographer, whose activities converge to the same end, namely, the geography of the present, moulded by the past and flowing into the future.

Now it may be objected that the above discussion is irrelevant or at least misguided, in that it rests on a misunderstanding of the true purpose and function of geography. The objector might contend that it forms a sufficient task for the geographer to discover and describe the present relationship between the physical and cultural elements of landscapes. The questions of origin and development, he might argue, should be left to the historian, since what matters above all to the geographer are the patterns and distributions of the present and human activities now operative rather than momentum generated in the past. Finally, he might conclude his attack by remarking that, even were it desirable and fitting that the geographer should acquaint himself with the history behind geography, to do so would impose an insufferable burden: like Atlas, the geographer has already the whole earth on his shoulders, yet he is asked to bear even more.

With these contentions we profoundly disagree on several

grounds. First, a geographical science which excludes all study of the processes which have moulded present areas alike in their human and physical aspects, is not only inevitably incomplete but also liable to fall into error. Secondly, such a science would bring to the study of the earth's surface so limited a perspective as to deprive it of much of its educational and philosophical possibilities. Thirdly, there is no reason to believe that, if the geographer neglects the study of past geographies, either the historian or any other specialist will necessarily undertake it. Finally, whilst it is admitted that the geographer has a vast field of study and one which requires acquaintance with the work of several other cognate sciences, we would argue the claim for a geography which is less extensive and more intensive in its objectives.

Admittedly, too, there is ample room for the specialization of studies within the geographical field, and only some students will wish or be equipped to make a special study of historical geography. But it is important that the student of geography should be aware of this aspect of his subject and assess its relevance. Desirably he should have some grasp of the geography of his own country in terms of its growth.

The reader will have grasped already why it is that the two contrasting views of the geographer's task which we have discussed above can be logically upheld. The reason is that any fact of human geography is capable of either a static or a dynamic explanation. Let us illustrate this statement.

The location of a town and its functions may be explained in terms of the particular physical factors, local and regional, which characterize its site and position and support its buildings and economy—this is the static explanation. They may be explained, too, by reference to the process of its actual foundation and growth—this is the genetic explanation. Similarly the industries of a town or area can be explained by reference to certain physical facilities, such as local coal, abundant soft or hard water, nearness to markets or to seaports. Such an explanation would exclude reference to their history—to the human initiative and enterprise which started the industries under conditions and for reasons which may no longer be self-evident. It will be agreed that the genetic explanation does not exclude the static, and vice versa; indeed, they are complementary.

Reflection will suggest the inadequacy and dangers of a solely static interpretation of contemporary phenomena. How could one account for the great city of Leningrad—the biggest industrial city of the world in so northerly a situation—unless something of its history was unravelled, notably the determination of an autocratic Tsar to capitalize a good geographical position accessible to the Baltic Sea despite, in contrast, the unpropitious character of the site? The recent growth of Magnitogorsk in the steppe south-east of the Ural Mountains is not explained solely by reference to the rich source of high quality iron ore provided by Magnet Mountain which dominates the town; reference has clearly to be made to Soviet policy and action which, in the last two decades, has created the chief steel-making city of Europe in a treeless area where, owing to the absence of wood for charcoal, it was believed down to the end of the 19th century that no iron and steel industry could ever be established.

We should be wrong, too, to suggest that the proximity of coal, as a source of power, had much to do with the original localization of the textile industries of Lancashire which illustrate what may be called either "historical momentum" or "geographical inertia." Actually these industries started long ago at a time when coal played no effective part either as a source of steam power to drive the looms or as a means of smelting the iron ore from which metal machinery was derived. Again, it might well be rash to conclude from the study of a map that a junction of roads explained the choice and importance of an urban site, unless the time sequence of the town and the roads had first been broadly established. Indeed, when the geographer with the map or country before him ceases to be merely descriptive and attempts genetic explanations, he is building hypotheses, to test which he must seek the facts behind the facts, which are historical in character, whether physical or human.

Before we try further to illustrate and justify our argument, let us note what were the main historical phases in the gradual formation of the countryside of southern Britain.

These significant phases cover little more than four thousand years, for the enormous preceding period of about half a million years of Old Stone Age culture had no geographical effects. During the former short period, so minute geologically, the physical changes which occurred, though small in scale measured by the geologist's yardstick, were not without importance in their human effect: the severance of Britain from the Continent and from Ireland, slight changes in the level of the land in relation to the sea, evidenced by the raised beaches of western and eastern Scotland and the submerged forests off Wales and eastern England, and the small-scale continual alternation of loss and gain of land along our coasts.

In contrast, this same period, historically a long and eventful one when compared with the recorded history of the Americas and of Australasia, witnessed a succession of human activity which left its mark on the countryside. Successive intruders entered Britain, occupied and colonized parts of its surface, and each in varying measure and in its own way modified that surface in ways still partially visible. We can liken these recurrent invasions and settlements to the seas which invaded the lands and contributed to the physical make-up of Britain at different periods of its geological history. These seas left deposits of sediments, part of which survives and remains exposed; similarly, the many immigrant peoples of Britain have left unequally residual traces of their presence.

Just as it is necessary to know the geological succession in order intelligently to approach the study of physical structure and land-forms, so it is necessary to be familiar with the main phases of the cultural succession. In contrast to the geological succession, which is applicable to the whole surface of the earth, that of human culture is particularized from place to place. The study of human history, unlike that of geology, admits of no general laws of widespread application.

There is a cultural history of Britain which follows broadly the lines of that of western Europe, but is different from that of China, Africa or the Americas in the local character, sequence and chronology of its cultural phenomena. Thus the historical geographer can best begin by studying his own country: in learning how one human geography has been moulded by cultural forces in the past, he will become at least aware of the nature of such processes in studying remoter settings. And in becoming for a time a "chorographer," that is a student of particular areas, rather than a geographer in the literal sense, he can enrich his understanding of countryside.

Let us tabulate the main stages in the cultural history of Britain. Prehistory, which derives its materials from archæology and physical anthropology, and history, which in the strict sense of the term is based on written records, can for the geographer's purposes be regarded as one. The importance of the various stages from the standpoint of geography, i.e., in the degree to which each has contributed to the present human geography, is very unequal, unequal too from one part of the country to another. The dates given below are susceptible to refinements unnecessary here: it is the sequence, rather than the exact limits of the periods, which matters.

THE MAIN CULTURAL PHASES OF BRITAIN

B.C. 2500	
	Neolithic Age, including Megalithic Period.
2000	
	Bronze Age.
500	
	Early Iron Age.
0	
	Roman Period.
A.D. 500	
	Anglo-Saxon and Scandinavian Colonization.
1000	
	Later Middle Ages.
1500	
	Agrarian, Commercial and Industrial Revolutions.
1950	

Surviving antiquities of the pre-Roman, i.e., prehistoric, period carry an air of detachment from the settlement and circulation of our present-day geography, although they are sufficiently abundant to catch the eye and to arouse the curiosity of the observant traveller. They consist for the most part of burial mounds—the Long Barrows of Neolithic age and the Round Barrows of mainly Bronze Age times which derive their name from the O. E. "beorg"=a hill and are described as "tumuli" on Ordnance Survey maps. To the barrows are added Earthen and Wooden Circles, Standing Stones, Stone Rows and Avenues, which express, grandly and timelessly, planned communal purposes, dimly though these now appear to us. Such ancient monuments were often chosen as bound marks, as land documents of the Anglo-Saxon period show, and the plough has to an extent difficult to measure—respected their claim to survival.

Of greater geographical interest, since they relate to settlement, are those evidences, which aerial photography is revealing increasingly, of the sites of ancient villages, of grain storage pits, of large defensible sites rather loosely known as "camps," of old trackways and even, towards the close of the prehistoric period, of arable field patterns.

Such antiquities are not uniformly distributed; rather they suggest rational selection of sites in relation to routes of entry and to geographical conditions of climate, soil, vegetation and drainage different from those which now obtain. The so-called megalithic monuments (long barrows, chambered cairns and dolmens) which owe their occurrence in Britain to peoples who entered the western, northern and south-western coastlands, are found, for example, in the lower littoral plains of Wales above the valley bottoms and below the forest levels and in England mainly between the Severn–Avon and the Dorset Downs on limestone plateaus, notably within the historic limits of Wessex.

The prehistoric "camps," enclosed defensive works of stone or earth, for the most part relate to the Celtic intruders of the Early Iron Age B., although a few, like Maiden Castle (Dorset) and Windmill Hill (Wiltshire), are known to have had a neolithic origin and successive stages of growth and occupation. These camps, which were probably as much hill villages as military refuges and strongholds—the "oppida" which resisted the Romans—occur frequently on the summits of our downlands, as also in Wales on sites ranging between 300 and 1,500 feet. Old trackways, too, the origin of which can be traced to pre-Roman times, although they bear later English names, can still be traced along the higher ground—witness the Icknield Way and the Pilgrims' Way which respectively pursue the escarpments of the Chiltern Plateau and the North Downs *en route* for the coasts of the Wash and eastern Kent, and the route which linked the prehistoric St. Albans via Wheathampstead to Colchester.

These roads, together with traces of upland cultivation in the Wessex downlands, in Fenland, Wharfedale, Montgomeryshire and elsewhere, have been associated with the Celts who inhabited Britain before the Roman conquest (A.D. 43), though the former may be more ancient in origin. In either case such vestiges provoke the interest not only of specialist students of our prehistory but also of geographers, since they form part of the contemporary scene. And no region of greater interest in the historical geography of England can be recommended than Wessex which, with its relative abundance of prehistoric monuments and features, had a centrality and importance in prehistoric geography which it retained down to Anglo-Saxon times.

Although the Romans did not contribute appreciably to the peopling of Britain—they were civilizing conquerors and rulers rather than settlers in scale—many visible geographical marks of their presence south of the Scottish Highlands have survived, thanks above all to their skill in engineering. The Ordnance Survey map of Roman Britain (second edition, 1928) indicates the variety of the evidence—roads, cities, "villas" (= farms), fortresses, forts, walls, etc., as well as its geographical distribution.

To the Romans was due the creation for the first time of well-built and planned towns, often, as at Canterbury, St. Albans and Cambridge, on or near pre-existing sites and, although relatively little survives today, e.g., sections of wall at Chester and York and the bath supplied by a hot spring at Bath, it was above and around the ruins of Roman cities that many of our present towns, notably London, subsequently grew. The Roman walls of the emperors Hadrian and Antoninus in north Britain, together with the "vallum" on the south side of Hadrian's wall which defined the first major political boundary within Britain, present more conspicuous records of the Roman period. The Roman roads, raised engineered highways, the symbol and vehicle of centralized government which permitted precociously the political unity of southern Britain, long played and still play a part in our geography. They do this not only as static patterns, as when stretches of these roads form parts of parish and county boundaries, but also dynamically, in affording means of local circulation and alignments for broad modern highways, for example, that from Canterbury via London to St. Albans and Dunstable.

As the film unravels, emphasis shifts to colonization, settlement, the fuller penetration of the countryside by farming peoples and to the break-up of the simple political pattern which the Romans, by their superior material culture and statecraft, had created.

Sea-borne, like all the intruders into Britain since the dawn of neolithic times, the Anglo-Saxon conquerors and colonists made landfall south of the Tyne on the eastern and southern coasts, and slowly pressed inland by way of the rivers and land routes such as the Icknield Way. Unlike the Romans, who conquered and occupied by means of centrally directed and well organized military forces, the Anglo-Saxons effected piece-meal local conquests, the objective of which was possession of land for agricultural settlement. Their settlement during the initial phase A.D. 450–650, when they were pagan and practised cremation burial, can be visualized from the distribution of their cemeteries, of place-name endings in "-ing" and "-ingham" and of heathen place-names attributable to this period.

Such evidence points to significant changes in the selection of sites and in the conquest of the soil. Farmers rather than townsmen and armed with a new agricultural technique embodied in their heavy plough team, more efficient than either the light wheelless *aratrum* or the heavier wheeled *caruca* of Roman Britain, the Anglo-Saxons avoided at first the Roman towns and highways and sought adventurously lowlands which had hitherto been avoided because of soil and drainage difficulties as well as the lighter, better drained, open lands that had long been used. In succeeding centuries the Anglo-Saxons consolidated their hold on Britain east of Cornwall and Wales and south of the Scottish Highlands, and established rural settlements widely as our present place-names testify, though some river names recall the earlier Celtic phase. Fig. 4 shows settlements—Anglo-Saxon and Scandinavian in origin—distributed along Humberside on the rising ground above flood level; southwards lie alluvial lands of deposition which have been reclaimed since the days of early settlement.

To the Anglo-Saxons also appears due the development later throughout an extensive interior belt of England of the new system of open, that is unhedged, fields, contrasting sharply with the "Celtic" field pattern of pre-Roman and Roman times.[1] They elaborated, too, a system of administrative units —counties, hundreds and parishes which, despite subsequent modification, still characterize our administrative-territorial organization. In the west midlands, forming the boundary of the Anglian kingdom of Mercia (=marchland) against the Welsh (an Anglo-Saxon appellation meaning "foreigners") and ostensibly the first demarcation of Wales from England, was constructed about A.D. 800 Offa's dyke, which can still be traced on the ground.

The Scandinavian Norsemen or Vikings—seamen, raiders, colonizers and state-builders, added yet another element to the peopling of the British Isles, making their way into eastern England north of the Wash (choosing the Humber as a major gateway) and into eastern, northern and western Scotland, thence to make land again in Cumberland, Westmorland, Lancashire, Wales and eastern Ireland. Canon Isaac Taylor in his *Words and Places* (1864) pioneered in this field of study, producing then a generalized coloured map of Celtic, Anglo-Saxon and Scandinavian place-names in the British Isles which, whatever its detailed imperfections, is still our best epitome of the extent and permeation of Scandinavian settlement.

[1] During the Roman period in Southern Britain two field systems obtained. One was the so-called Celtic field system, associated mainly with native villages, and the other was an open field system employed on Romano-British farms (*villæ*). There is no evidence to suggest that the medieval open field system derived from this latter system.

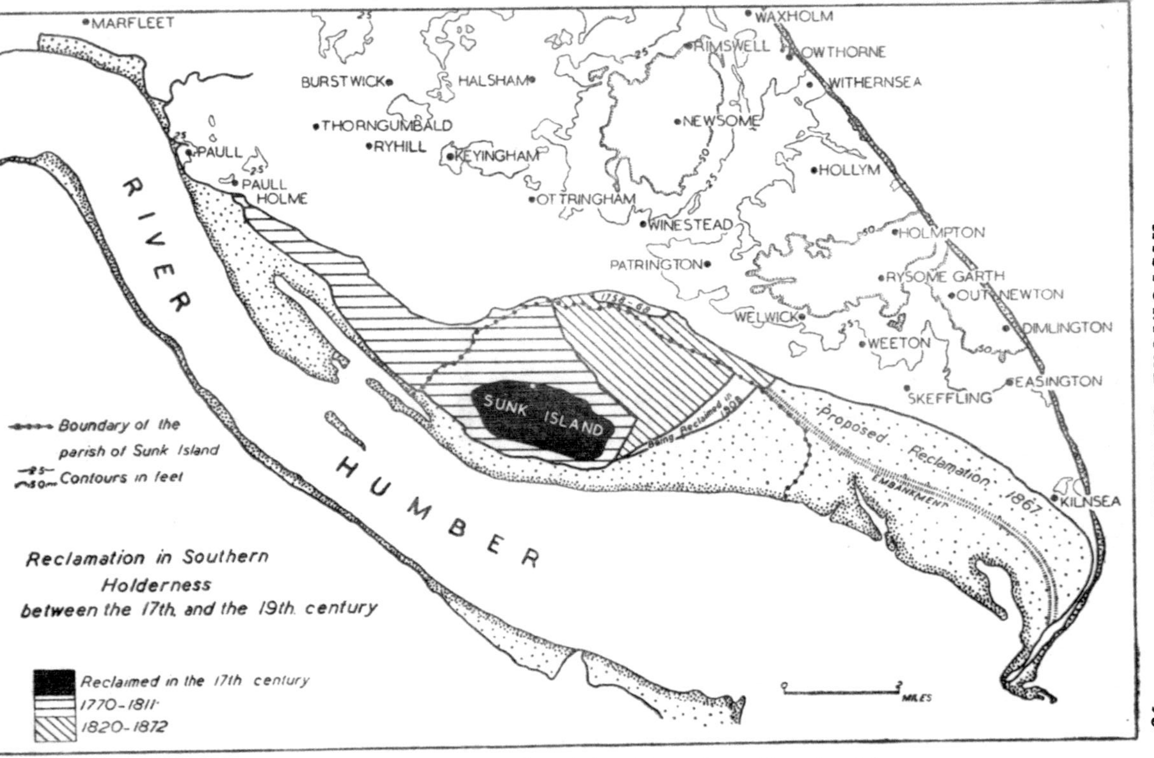

Fig. 4. The Settlements and reclaimed Land on Humberside.

This settlement was substantial, as it were filling the gaps left between the widely spaced Anglo-Saxon sites, notably in our northern counties, measurably as far south as East Anglia and Bedfordshire, but marginally in the western lands—eastern Ireland, western Scotland and Wales, where Celtic-speaking peoples, originally established in the Early Iron Age or yet earlier, defended in relative remoteness their traditional culture. The Scandinavians, too, like the Anglo-Saxons by whom they were appeased and with whom they fought and shared the political control of southern Britain, helped to form our pattern of counties in the midlands in relation to army districts allocated to fortified "burhs," e.g., at Leicester, Northampton, Nottingham and Bedford, and created such administrative anomalies as the "soke" of Peterborough and the "ridings" of Yorkshire.

The Domesday Survey of England, drawn up by King William I in A.D. 1086 and now being made to yield materials for a geographical picture of England—all save the four northern counties was covered by the survey—shows that for some parts of the English Plain the rural settlement sites of today were then almost wholly established. It points to a zone of agricultural country in south-central England, from East Anglia to the Bristol Channel, as the most densely settled and cultivated part, though it reveals no less clearly tracts still largely negative, where settlement had made little intrusion into the forests, as in the English Midlands and the central Weald, and as in the marshlands of Somerset and Holderness.

Although the Normans, like the Romans, did not effect any considerable colonization, they brought the same will to achieve political unification and showed a similar skill in building. Hence the creation of marcher lordships on the frontiers of the north and west against the Scots and Welsh respectively, the analogous political status of Fenland since its physical character facilitated resistance to royal power in the midst of the English Plain, and the construction of castles in stone some of which—unlike the wooden and earthen structures of Anglo-Saxons and Danes—have partially survived. Under the ægis of Norman kings and their successors and that of the Church together with its monastic orders, new lands were reclaimed in the forests and marshes (although much land of varied character was stabilized as royal forest), and cathedrals, abbeys and parish churches were built, many of which with subsequent accretions survive to permit us to recapture the architectural achievement of those days.

England in the later Middle Ages, notably between A.D. 1300 and 1500, presents a highly individualized human geography in which the trade and industry of the towns and the open-field agriculture of the countryside were over wide areas characteristic features of the economy.

It must suffice here to underline some aspects of this phase which bear closely on the formation of our present geography. We may note then how, increasingly after the Norman conquest, London, associated with the royal capital at Westminster and later as the permanent seat of the royal courts of law, developed also its functions as a trade, port and route centre and asserted the dominance which it thereafter consolidated as the capital and route focus of England. This development marked a shift in the centre of gravity of southern Britain in that, although London had appeared prematurely as the chief road centre of Roman Britain, the primary areas of settlement and political power at other times lay elsewhere: in the St. Albans–Colchester area before the Roman conquest; in the upper Thames between the Chalk plateaus of the Chilterns and Wessex in the early Anglo-Saxon phase; and subsequently, in the Wessex downlands, as in prehistory. where Winchester provided a capital.

No less characteristic of the later Middle Ages was the rise of towns, differentiated from the feudalized countryside by their enjoyment of civic powers of self-government and concentrating in their guilds, markets and fairs substantial industrial and mercantile activities. Some of these grew up around the historic sites of Roman cities, as at Lincoln, York and London, or of Danish "burhs," as at Nottingham Derby, Stamford and Northampton; others were the "chesters" —Gloucester, Worcester, Winchester, Manchester, Exeter, etc., by which our ancestors described former Roman sites though they were not "castra" (fortresses) in the strict sense. Others under royal patronage developed urban functions though at the time of Domesday Book they constituted, or formed part of, largely rural manors, for example, the seaports of Bristol, Southampton and Kingston-upon-Hull, which received its charter from King Edward I., and the inland cities of Norwich and Coventry. In effect, this rise of towns associated in large degree successively with the trade in, and working of, wool, a major source of Britain's wealth in late medieval times, added a new feature to the countryside—relatively numerous and populous industrial settlements with large churches contrasting with the more widely spaced official and largely non-industrial centres of earlier periods.

Increasing movement of people—the Court, merchants, judges, pilgrims, armies and, within a small orbit, peasants, and the commodities of interregional and international trade—stone, timber, wool, fish, salt, lead and grain—indicate the availability of land and water routes, somewhat limited in character but by no means inadequate to the needs of the time. The roads carried the major traffic: some were still aligned along Roman roads which, however, had greatly deteriorated through ignorance of technique, neglect and the plundering of their stone; others, notably that which linked Southampton and Northampton and helps to explain this association of place-names, are post-Roman roads, independent of London, and are shown clearly on the so-called Gough map, which was drawn *c.* 1350 on a sheep skin preserved at the Bodleian and reproduced by the Ordnance Survey. Much attention was paid to bridge-building. The rivers, the greater ones, notably the Thames, Severn and Yorkshire Ouse, and many smaller ones, the Witham, the Bristol Avon, the Cambridgeshire Ouse, and the Don, were called into service to relate port and hinterland, whilst the seas already provided lanes for coastwise traffic. Foreign trade made use of all the British coasts—London, Lynn and Hull, the chief east coast ports, as well as Southampton and Bristol: by the 15th century English ships were sailing not only to Ireland and across the Channel, but to the Baltic, Mediterranean and Iceland.

The rural landscape presented transitional geographical features during the pre-Industrial Revolution period. Population had been steadily growing: estimates for England

and Wales indicate barely one million in Roman times, 1.5 millions at the time of Domesday Book, 5.5–6 millions in 1700, reaching 9.2 at the first census of 1801. This was a largely rural population engaged in the production of foodstuffs and raw materials of industry. In the midland belt of England, where the open field system was most typically established, the process of enclosure of arable fields, at first primarily to facilitate sheep rearing and later to permit specialized capitalist farming for food, was changing the face of the country between the late 15th and the early 19th century. The early phase of enclosure for sheep entailed the abandonment of many nucleated villages, and the study of these lost villages is now taking place.

The chequer-board of irregularly shaped hedged fields, so characteristic of our present countryside, was the geographical reflection of these great social and economic changes. The west of Britain, which initial Anglo-Saxon settlement failed to penetrate, received the open field system belatedly, e.g., in and after the Norman period in South Wales, and at the end of the Middle Ages enclosed fields were already characteristic of limited arable tracts. In south-eastern England—in Kent and Suffolk—a field system, based on small enclosed arable fields, not dissimilar from that which appeared in Celtic and Roman Britain, persisted, genetically alien to the open field system. Side by side with the later phase of the enclosure movement and again marking new features on the face of the country came the construction, especially in the 18th century, of stately country mansions—classical styles were favoured—the laying out of parks and ornamental plantations for which many new tree species were introduced.

The industrialization of Britain, effected by what Arnold Toynbee first called "the Industrial Revolution of the Eighteenth Century"—though indeed these processes of change had their roots in earlier times and continued unabated into the 19th and 20th centuries—produced by stages a new geography of landscape and social relationships.

Enough has been said to suggest the many successive human geographies of the past, which the historical geographer can attempt to reconstruct, and strands from which are interwoven in our contemporary geography. We are not here concerned with the means and technique by which such work is undertaken. It is scarcely the easy task implied by a former Director-General of the Ordnance Survey who said in a light mood that "with a few dates, and with a good general idea of the main features of the geography, we can picture the past clearly, both in time and space." We are not arguing here that all students of geography should become fully trained historical geographers. Rather, we submit that the student of present-day geography should be cognizant of the geographical legacy of the past if he is to interpret with understanding the world around him.

A few illustrations are selected to emphasize how parts of our geography are the handwriting of history. The county boundaries of England, defined for the most part in the Anglo-Saxon–Scandinavian period, enclosed areas geographically symmetrical, for example, Kent, Durham, Cornwall and Herefordshire, and others until their recent readjustment, geographically preposterous, such as Worcestershire, Gloucestershire and Warwick (Fig. 5). The first-named county contained "peninsulas" jutting into Gloucestershire and five "islands" set within Gloucestershire, Warwick and Oxfordshire. No detailed study of the administrative map would have elicited the explanation of these anomalies; they are, however, historically explicable, as Sir Charles Oman showed.

It appears that when the county boundaries of Worcestershire and Gloucestershire were determined, long before Domesday Book and perhaps in the reign of King Ethelred the Redeless, with few exceptions these islands and peninsulas belonged to influential monastic houses which were able to ensure that they were placed in the same shire (i.e., in Worcester or Gloucester) as their central foundation. The separation of Furness from the rest of Lancashire as a result of the intrusion of Westmorland has doubtless also its historical story. Of smaller interest are such oddities as the south-western boundary of Cheshire between Chester and the estuary of the Dee, which appears unaccountably to avoid the course of the Dee until it is noted that canalization of the Dee has withdrawn it there from the boundary with which it was originally associated.

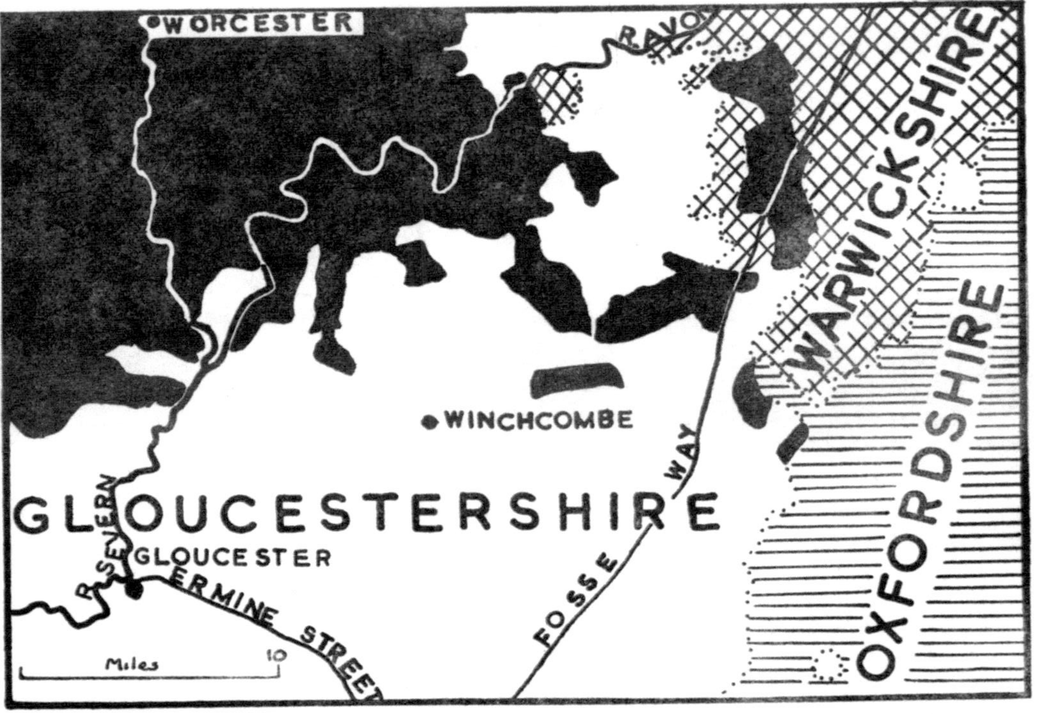

Fig. 5. The County Area of Worcestershire (shown black), before the Revision of its Boundaries by the Gloucestershire, Warwickshire and Worcestershire Act of 1931.

The three successive courses of the lower Don, all of which present visible features today. provide instances of past geographies mingled with that of the present (Fig. 6). The original historical course of the river, now an abandoned waterway, meandered north-eastwards to enter the Trent near its confluence with the Ouse. Along it, as place-names testify, were established in turn Anglo-Saxon and Scandinavian settlements by immigrants who doubtless used the then open waterway which still forms part of the common boundary of Lincolnshire and Yorkshire. This earliest historical channel of the lower Don was sanded up in the 14th century by Humber silt during a period of exceptionally stormy conditions, and in the 15th century the Don was making its way seawards by way of the "Dike" which carried its waters to the Aire, east of Snaith, *en route* for the Ouse and Humber. Finally, the present outlet of the Don to the Ouse at Goole, known as the Dutch river, was cut in the 17th century.

Again, to present a different illustration, are we sufficiently aware that Britain has no aboriginal scenery, and that what we are inclined to call "natural" landscape or vegetation is more fittingly described as "wild nature," so great and continual have been the changes effected or initiated by its inhabitants during the last few thousand years? The high moors of northern England were originally covered by forests which have been destroyed by man. Bracken, which clothes many of our hill-sides, is the natural response to the clearance of woodland, and thus not aboriginal. The downlands of southern England were lightly covered with beech, box, juniper and thorn, and more thickly wooded where heavy soil deposits overlay the Chalk. The New Forest, in so far as it is woodland rather than heath, owes this character to the plantations of recent centuries, whilst Breckland (West Norfolk), now undergoing afforestation, was the nearest thing in Britain to true sandy steppe and as such was much sought for settlement in prehistoric times.[1] And how few of our familiar tree species,

[1] The vegetation story of Breckland is not simple. The study of pollen analysis indicates that, when Neolithic man arrived, the area was covered by oak forest; its subsequent reversion to heath was the result of forest clearance by Neolithic man. See G. Clark, "Forest Clearance and Prehistoric Farming," *The Economic History Review*, XVII, No. 1 (1947).

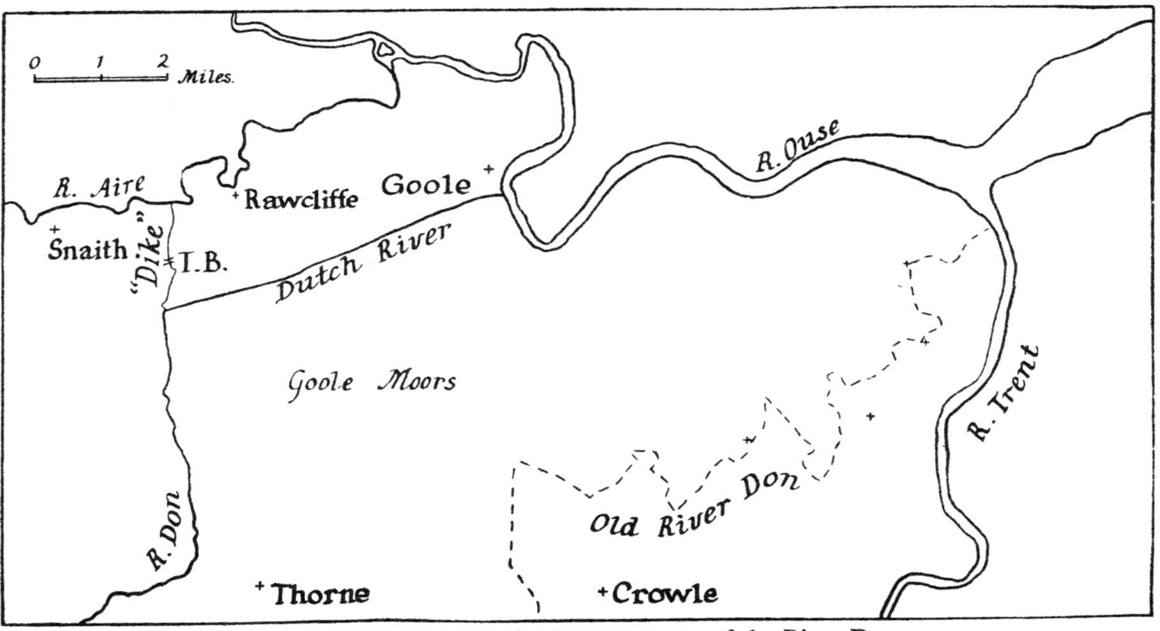

Fig. 6. The three successive lower courses of the River Don.

save the oak, ash and in Scotland the Scots pine, are known to be indigenous?

Enough has been said to demonstrate that the geographer in his study of the static and dynamic aspects of the contemporary countryside is concerned with the dimension "time," alike in its relation to physical and to human history. It is all to the good that, in turning away for a time from the whole world, he becomes what would once have been called a "chorographer," interested in some part of the whole and not least in his country of residence since, as an able broadcaster used to exhort us, to dig ever more deeply is the prerequisite of an abundant crop.

Nor need the geographer fear that he will find, as did a British Ambassador in Moscow during the 1939–45 war, that he has been digging in error the plot of another since, in so far as historical questions arise in the course of his study of contemporary geography, it is clearly his business to pursue and try to solve them.

CHAPTER VI

ECONOMIC GEOGRAPHY

Say Lords, should not our thoughts be first of commerce?
My Lord Bishop, you would recommend us agriculture?

WILLIAM BLAKE, *King Edward the Third.*

ECONOMIC geography so-called has its place in syllabuses of geography and is not unfamiliar to the layman. At first sight its content and purpose are self-evident, once it is grasped that geography is concerned not merely with the position of places and the shapes of countries, but also with man and his works in so far as they are related in place. Production and trade in all their manifold aspects are not only in some sense facts of geography, but they reflect in striking fashion the geographical differentiation of the surface of the earth. Here, by common consent, is a branch of our subject both important and useful, and the existence of a wide and growing range of texts, bearing the comprehensive title "economic geography," would appear to characterize clearly its field. Let us glance briefly at some of these to see what is currently accepted before turning to certain difficulties and ambiguities which invite discussion, since these are not merely points of nomenclature, but must affect our thinking over the whole field of geography.

The term "economic geography" appears to have been first used in 1882 by the German Gotz to distinguish his work from "Commercial Geography" which had the earlier start. This latter raises no special problems: its purpose is frankly practical rather than philosophical, scientific or educational in the wider sense. Commercial geography presents summaries, periodically brought up to date, of the production and trade of the principal commodities of the world, set against its variegated geographical background. Rarely does it essay more than factual statements, replete with statistics, about the location, scale and processes involved in the production and exchange of commodities; rarely does it attempt more than to note in passing the economic and political considerations which have a substantial place in any explanation of the facts set forth. It is primarily a work of reference, supported at best by a systematic appraisal of the major facts of General World Geography.

At times the attempt is made to set side by side the geographical conditions, which appear to indicate production and trade of a certain kind, and the economic and political conditions which modify these expectations. In this context reference may be made to the late Dr. M. I. Newbigin's *Commercial Geography*, where the reader is at least made aware of the inconstant politico-economic factor which intrudes markedly into the geography of agriculture, of industry and of trade.

For Gotz economic geography was commercial geography causally treated: its objective was a scientific study of world areas in their direct influence upon the production of goods. Here was defined a path along which many economic geographers have trod, seeking above all to lay bare the influence of physical factors on the occupations, products and more generally the life of peoples established in the different regions of the world. Here was a task pre-eminently geographical, directly related to, and overlapping, the field of General Geography, which attempts to establish for the world as a whole the nature and rationale of its regional differentiation. Here, too, was a task which had something in common with that of the economist, whose subject matter was no less than the wealth of nations.

But it is not apparent in the many economic geographies which are offered as college texts that a really hyphenated subject has been created in the sense that their authors are cognizant of the economist's interest in geography, his concepts and his laws, or are seeking to answer in the light of their own expertise questions which he might raise, with reference, for example, to the theories of comparative costs and the division of labour. Nor for that matter is it evident that economists as a whole find the need of support in their study of the material systematized by the economic geographer. Of the latest works of Pigou and Keynes it was stated in 1936 that they were "nearly bare of *geographical facts*" and, if the words of the father of economics still represent orthodox dogma, geography is no desideratum in the study of economics. For Adam Smith in the first page of *The Wealth of Nations* lays it down firmly that:

> "Whatever be the soil, climate, or extent of territory of any particular nation, the abundance or scantiness of its annual supply (*sc*. of the necessaries and conveniences of life) must, in that particular situation, depend upon . . . two circumstances:"

—namely the skill, dexterity and judgment with which its labour is generally applied and the proportion between the number employed in useful labour and of those who are not so employed. And, although *The Wealth of Nations* has been described as a "systematic induction from history and geography" which contains "more geographical references in proportion to its bulk than any later important work (*sc*. oi economics)", there are grounds for believing that both by geographers and by economists the common borderland of their studies awaits fuller exploration.

But returning to the economic geographies, let us note what they set out to do and their limited success. For the most part the work of geographers, their objectives and treatment are various. For one standard text "the economic activities of Man in their physical and cultural setting"—no less than this—is the object of study. Others in varying degree seek to assess the physical and cultural factors in the economic activity of peoples within selected regional frames. Treatment by major commodities forms commonly a part of such general works. Some geographers even see the affinities of economic geography less with the science of economics than with historical geography: economic geography, concerned with the present, forms together with historical geography, the content of human geography. To this others would react, asserting that economic geography is only part of the wider field of "human" geography, and that the economic aspect of human geography has been so exaggerated as to obscure and to subordinate to itself the field appropriately called social geography.

Economic and Social Geography, by E. Huntington, F. E. Williams and S. van Valkenburg, may be looked at more closely as a good specimen of such college texts. These authors indicate by reference to natural rubber the kind of questions which arise and how they can be tackled. The *hevea* tree has certain specific climatic requirements which limit its possible growth to low altitudes of the hot and wet lands near the Equator, but it is immediately evident that its growth is in fact restricted to only a very small part of such lands. To some extent other physical factors, notably relief and soil, serve to explain why much of the climatically suitable land cannot in fact be used, whilst another geographical factor, position in relation to the sea and to the main frequented seaways of the world, intrudes to underline the advantages of certain areas of potential rubber cultivation, notably Malaya and the East Indies, since rubber is a commodity of world trade and sea transport is cheap, especially on a sea thoroughfare such as East Indian waters.

At this stage the argument begins to take into further account economic as distinct from geographical concepts. Natural rubber, the production of which is geographically limited, is in demand, notably by those countries of North America and Western Europe, where in this century it has become an essential element in industries manufacturing automobiles, bicycles, clothing, footwear, etc. Changes in demand, by their reaction on price, have geographical effects in stimulating the extension or contraction of plantations, the cultivation or use of wild rubber and so on. Another economic factor is labour supply, and the superior position in respect of both numbers and quality of native labour of south-east Asia as compared with tropical Africa and Brazil is clearly one factor behind the present location of rubber cultivation. Similarly, the competition of rubber and rice for the land to which each is equally suited raises an economic issue and invites an economic solution—i.e., which is the more profitable use of limited available land?

At this stage political considerations too enter, for Governments may decide not to allow price to be the sole arbiter and may prescribe the proportionate allotment to rice and to rubber. No less is it relevant to note such political facts as the instability of government in Brazil and the former colonial status of south-east Asia. Clearly the policies of Great Britain and the Netherlands were directed to develop under the plantation system and by the application of capital the production of this valuable counter in world trade. The successful application of chemistry and electric power to the production of a variety of synthetic rubbers in Nazi Germany, the U.S.S.R. and the United States, for reasons essentially political and regardless of economic criteria, raises yet another complication bearing on the problem of the economic geography of rubber.

Such discussion, supported by accurate information and illustrated by maps and diagrams, of problems which concern both geography (i.e., Place factors) and applied economics (i.e., the production and exchange of commodities), although it ranges widely, claims to fall within the scope and purpose of economic geography. Such problems are admittedly intricate enough, and call for careful analysis and critical handling. Their exposition involves considerations ranging over the whole field of physical and human geography, some knowledge of recent economic history and politics, and an awareness of at least some of the elementary principles of economic science.

They serve on the one hand to illustrate, for the student of geography, the complexity and variability of the economic factor as it impinges on the use of land and helps to determine its patterns and the flow of its products. They serve, too, to provide that realistic physical basis to the theoretical superstructure with which the "pure" economist is concerned, and to provide him with illustrative material to illumine his abstract expositions. Logically, it would appear, that fruitful collaboration might be effected between economic geographers and workers in "descriptive" or "applied" economics, i.e., those concerned with finite particular problems in an areal setting, on such problems of common interest as the location of industry, the division of labour and comparative costs.

But it must be admitted that economic geography as a specialist branch of geography has not yet advanced so far as to have established its position beyond question and to have differentiated itself clearly from other branches of the whole subject. In the text to which we have alluded above, a large part is devoted to climatic and physiographic matter which is common to General World Geography, and the very balance of the book tends to over-emphasize climate in particular, the more so since its readers, being mainly students of geography, are little aware of the relevant principles of economics for which space might have been profitably found. For it is here that the main difficulty arises, to which we alluded on page 103 above and to which we now return.

What significance attaches the adjective "economic" in the above context? We speak also of Economic History, Economic Geology, Economic Botany and Economic Entomology. To the first we shall refer again, but it will at once be argued in the other cases that the word "economic" carries a perfectly definite and limited meaning. It connotes those parts of pure sciences which are concerned with useful applications. Geology studies the constitution and structure of the crust of the earth, and the physical history deduced therefrom. It thus collects facts and develops theories immediately relevant to the search for, and exploitation of, coal, metallic ores, petroleum, underground water, etc. The particular study of the geological problems connected with the occurrence of these substances is thus aptly named, in that they are economically significant. Similarly, with all the "pure" sciences there is an applied branch, the relation of which to the main subject is not in doubt. It is noticeable, however, that we rarely if ever speak of economic chemistry. Though the pure and applied aspects of chemical science can be regarded as in large measure distinct, the latter is so vast in scope and so important that the common sense of everyday usage refuses to regard it a mere subdivision of a "pure" whole.

It will readily be agreed that when we use the term "economic geography" we are not using the adjective in the above sense. It is possible to conceive of an applied geography—the application of knowledge of the land surface, atmosphere and oceans, such as is made by the marine or aerial navigator, the land surveyor and many others. Nor can we deny that the subject matter of economic geography is useful in the narrow or applied sense. Many clerical and commercial workers are required to study it, because its facts and principles form part of the minimum equipment for efficient and intelligent work. Nevertheless, a wider meaning of "economic" is clearly implicit in the writings of geographers, and we perceive part, at least, of the reason for this when we recall that economic geography deals with some of the same concepts and topics as economics itself.

It would be regarded as unreasonable if an economic geologist were expected to be proficient in the principles of economics. It is true that the exploitation of mineral bodies raises purely economic questions, and the geologist might be none the worse for a knowledge of these. In practice, however, it is usual to treat the questions as separate and successive: the economic questions can hardly arise until the geologist's problems are largely solved. But we instinctively agree that it is by no means obvious that the economic geographer can ignore the findings of economics, and we must enquire therefore how far our instinct is sound.

We must take note at the outset of the marked differences of emphasis among the declared objectives of economists themselves. While few would deny that the central themes of economics are well known in the world of learning, there has been little agreement about the precise content and aim of the study. In Marshall's classic, *The Principles of Economics*, it was defined as "the study of men, earning a living," but it has also been defined as the explanation of the causes of material welfare and as the study of the methods by which men co-operate to meet their material needs. Other definitions appear more limited inasmuch as they refer to the phenomena of "price" or "money" in the context of material welfare.

It is undoubtedly true that the "material welfare" definition is at once too wide and too narrow, and quite untenable positions arise from any attempt to maintain it literally. It is not the whole of man's material welfare that comes within the purview of economics; nor can we safely exclude from the field certain definitely non-material aspects of his welfare. Not only goods but services which enter into the circle of exchange and are bought and sold, form part of the phenomena to which economic generalizations apply. The supply of, and demand, for ministers of religion, psycho-analysts and artists, for example, are all subject to essentially economic laws, no less than those of and for coal, rubber and petroleum.

In a lucid analysis, Professor L. Robbins, taking note of these evident considerations, re-defined economics as "the study of the forms assumed by human behaviour in the disposal of scarce means between alternative uses." Although this may appear a rather abstract statement, reflection assures us that it is a true and accurate rendering of the case. "Every act," he wrote, "which involves time and scarce means for the achievement of one end involves the relinquishment of their use for the achievement of another. It has an economic aspect." The individual, let us say, requires, *inter alia*, food, clothes, tobacco, books and visits to the theatre and to football matches. Economics studies the complex consequences of the choices made, not by one but by many individuals in attempting a quantitative sharing of means between such diverse ends. Each one "economizes" in the sense of making limited means go as far as they will, distributed between a multiplicity of ends. Two other conclusions of Robbins's analysis are relevant to our discussion. He claimed that the above definition was comprehensive in that it covered the case of the isolated "economic man," of socialist and communist forms of political organization as well as the individualistic exchange economy which prevails (if incompletely) in Europe and America. He admits, however, that it is chiefly in the last named that the economist finds the complexities which invite and require his more profound effort and analysis. Further, in the light of his definition he is able to indicate the appropriate rôles of Economic History and what has been called Descriptive Economics.

Economic History becomes in his view the study of the instances in which the relationship between means and ends has manifested itself through time. Similarly Descriptive

Economics, the study of recent or current individual instances of scarcity relationships is, as it were, the economic history of the present. Robbins points out with a certain asperity the limitations of such particular studies: they are random samples which do not, in fact, and cannot, by their very nature, give rise to sound generalizations of universal applicability. If we wish to assess the quantitative effect of some arbitrary fixing of the price of a commodity, it will avail us hardly at all that twenty years before, under entirely different conditions of supply and demand, the "elasticity of demand" was found to have a certain value. In the interval the whole economic and political environment may, indeed must, have changed. New substitutes now available, and new habits and fancies control the consumers' choice. We can in short derive no quantitative guidance from a study of twenty-year-old problems which must stand simply as a piece of economic history.

Robbins writes with a notable lack of enthusiasm about studies of this kind and deplores also the intrusion into economics of considerations concerned with the technique of production and distribution themselves—the study of ends and means as such. These he would regard as assumed data in the economic field and he envisages the further fruitful development of his subject not from the study of random samples, but from the careful refinement and elaboration of the tools of pure analysis. It is very relevant also to our discussion to note his comments on certain older treatments of the economic theme. He contrasts the "masterly sweep" of Marshall's Book V, dealing with a strictly economic problem, with the irrelevant technological intrusions in Book IV, e.g., "spineless platitudes about manures." He comments, too, on the insufferable dreariness and mediocrity of those earlier accounts which deal with industrial "forms," the supposed influence of national characteristics upon productivity, and the localization of industry.

It would be an impertinence for those not claiming the title of economist to criticize these views. They have, indeed, been criticized, as for example by Barbara Wootton, who accused the theoretical economist of living in an entirely unreal world, dominated by a perfect market mechanism, itself an almost ludicrous travesty of the market conditions which actually exist. She would prefer to define the economist as a student of social welfare, encouraging him to make realistic and institutional studies, to form opinion and to guide research for social betterment. It is no part of our business and in no sense within our competence to take sides in this ardent cleavage of opinion. It is probable that most students of the physical and natural sciences will feel considerable sympathy with the standpoint maintained by Professor Robbins. For them it is comparatively easy to segregate "free" elements and temporarily to hold variable factors constant. Thus they can build up a controlled laboratory technique which is denied to the economist, and they cannot fail to appreciate the economist's difficulties.

The chemist has in his own garden an example of applied chemistry at work; the difficulties of deducing the principles of pure chemistry from study in the garden are evident, and no chemist will fail to prefer the controlled facilities of his laboratory. But this is not to say that the principles of chemistry could *not* be evolved by a combination of observation in the garden and reflection in the study, if circumstances thus confined his effort. It is far from true that the whole structure of physical science is based upon experiment; in ever increasing degree pure mathematical analysis plays its part and observation often poses a problem for analysis either to solve or at least to illuminate. Certain it is that a very close analogy exists between the position of meteorology and that of economics. No proliferation of local and temporal "realistic" studies could solve the problems of the hydro-dynamics of the atmosphere, and several generations of such work proved abortive in producing anything but samples and most insecure "principles" of weather forecasting. Recent improvements are due in large measure to the application of an analytical method.

Our discussion of the nature and aims of economics, brief and inadequate though it is, may seem to have carried us far from the field of economic geography. We undertook it, however, in order to see how far any effective liaison between geography and economics was to be regarded as natural and proper. If we turn to the declarations of self-declared economic geographers, the necessity of enquiry is at once apparent. As we have already noted, one favoured definition of economic geography, at least in British and American circles, represents it as the study of the influence exerted on the economic activities of man by his physical environment.

In the light of the foregoing discussion this appears for a number of reasons far too wide a claim. It tacitly assumes the unknowable division into "economic" and "non-economic" things which enters into the narrower concept of material welfare. We have seen that many important scarcity relationships, subject to economic analysis, are not concerned with "goods" in the sense in which that term would be used by a railway clerk. The writers of this book, together with many of its probable readers, earn their living by what Adam Smith would have classed as non-productive labour. Only occasionally do we produce a commodity which we properly regard as "vendible," and we should be hard put to it to earn a living thereby. Is this then our only economic activity? Plainly not, since the valuation of our "non-productive" services is subject to the law of supply and demand. How then can we trace the relationship between these activities and our physical environment? Shall we say that an accident long ago in the juxtaposition of marshy valley floor and gravel terraces fixed the site of a future national and imperial capital in which there comes to be need and occasion for our services? Plainly in such an argument the chain of causation is drawn absurdly fine and long, and we should gain little intellectual stimulus or academic credit from such arguments.

It is probable that the reader of pragmatic tastes may well feel impatient at the seeming pedantry of the above argument. Is it not clear, he may maintain, that there is a tacit common sense limitation in the above definition, and that no one is, in fact, deceived as to the content of the subject? Let us take him at his word and admit that in fact the influence of physical environment is chiefly, though not exclusively, exerted upon the production of vendible commodities. What then is to be said about the many important towns, the populations of which find their major activity in administration in any of its aspects? Plainly there is a geographical aspect in the siting and functioning of such towns. Whether we regard their activities as economic or not, are the geographical aspects of such towns to be excluded from the purview of economic geography?

The conclusion of the argument now begins to appear. We set ourselves no clearly determinate problem in seeking to discover the relationship between physical environment and man's *economic* activities. Whether we accept the wider or narrower connotation of "economic," man's economic activities represent the vastly preponderant share of his total activities in the field of geography. If we insist on the above definition, we are simply defining as "economic" three-quarters or more of human geography and using a high-sounding adjective to create a seeming special division of our common subject matter. This, in our view, is precisely what has been done in many quarters, and for reasons not wholly sound or praiseworthy. In some confused way the prefix "economic" may have seemed to render the subject useful and important. The service and importance of geography can be demonstrated on better grounds than this, while leaving us free to employ our overworked adjective in a much more precise and serviceable fashion.

In the first place, it is clear that in the purely geographical study of a region or of the world as a whole, we perceive the influence of certain purely economic factors which react with physical, political and other factors in shaping the distributions and patterns shown on our maps and in the cultural landscape as seen on the ground. In such cases it is necessary to evaluate the economic together with the other factors in seeking to comprehend what the map and the ground present to us.

A simple example will make the matter clear. Southeastern England presents diverse local combinations of soil and climate which vary in their suitability for the growth of wheat, fruit and dairy produce. Up to a point, the sharing of the area between these three forms of production can be explained on physical grounds. The large-scale geological map of the area and to a less extent maps of climatic distributions, provide the bases for a number of correlations. *Ceteris paribus* we expect to find the grassland for dairying on the heavy clays, wheat and associated crops on soils of slightly lighter character, and the lightest soils of all unutilized agriculturally. The patterns thus prescribed solely by geological conditions find a partial realization in the area and some departures from the expected patterns can be explained in climatic terms. Thus, large parts of north Kent are comparable in soil conditions with the loam area of the Sussex Levels. The fact that the latter area engages in mixed farming, whilst much of the former is devoted to specialized fruit farming, is due partly at least to the slightly but significantly heavier rainfall of the Sussex coastlands.

But many other departures from the geological patterns cannot be explained in this way, nor in any other way which involves purely physical considerations. Thus the dairying belt on the London Clay in south-west Essex does not extend into the similarly constituted tract of eastern Essex. Confronted by such a fact, trailing shreds of determinism from an all but forgotten past, the geographer is sometimes tempted to invoke occult physical differences. Such differences, occult in the sense that they escape his first general scrutiny, undoubtedly exist in some places, but they do not afford the only or necessarily the chief differential in distributions.

In the instance before us we have concealed under *ceteris paribus* the prime economic fact of the great consuming market of London. In respect of perishable or quickly consumed commodities, other things are not equal, whilst distance from London varies. It is indeed a principle at once simple and very important that large towns impose a concentric pattern on land utilization in respect of dairy farming, market garden produce and the like. Similarly the differences in the character of fruit farming within the Kent area are related to distance from London. In the light of this example it will be evident that numerous distributions and the landscape features which they represent are powerfully affected by the economic factor of distance or, more fundamentally, of accessibility. For the economist it is enough to say, for instance, that limited depth of water or limited quay-side accommodation renders the handling of coal "uneconomic" at a certain small port. The geographer, however, is interested in the limiting factors themselves, and it is his business to discover just why and how they have helped or hindered "development" and thus made his region different from what it might otherwise have been. In a word, he takes into account the economic factor called into operation by variations in distance and accessibility.

In thus reckoning with the economic factors our geographer, let it be noted, has not converted himself into an economic geographer in any specialized sense. He is still just a normal and properly conducted "geographer" pursuing the ordinary analytical method of the subject. Nor must he seek to dignify or canonize the simpler economic principles which figure among his tools. In very large measure they are self-evident and amount to little more than pompous banalities when couched in abstract form. The interest and complexity of the matter in the field of geography lies in the fact that each case differs in detail from the others: the economic factor impinges in different fashion and in different degree upon the complex of other factors, and in the disentangling of the factors lies a worth-while intellectual task.

Let us now consider a quite different aspect of the relationship between things geographic and economic. The geographer, we repeat, cannot but be concerned with economic activity, since it is undoubtedly the single most powerful reagent in "developing" his cartographical "plate." Again it matters little whether we take the wider or the narrower sense of "economic." Social and political activity can and do leave their geographical mark on the map and the ground, but economic activity in some of its aspects inevitably must. A large part of the occupied areas of the earth is devoted to the production of the stuff of "material welfare," and the machinery of that production, in the widest sense, forms a preponderant element in visible or material culture.

The modern geographer will wisely scout the rash generalizations of former days which sought to relate the ethical and political ideas of a society to its physical environment. He will admit that in certain notable cases the inverse relation is true:

the social and political ideas affect the detail of the geographical environment. And in the field of economic activity the mutual relationship of Man and Land is evidently and inevitably closer. Whether we study the West Riding of Yorkshire, Sierra Leone, Cyprus or Samoa, the pattern of human geography will reflect in large and probably overwhelming degree the nature of prevailing economic activity.

This self-evident fact again seems to be the root cause of muddled thinking by many beginners in the subject. The fact that the activities written on the map are in the common sense economic does not demand that our study be labelled economic geography. In general it has as much or as little claim to this title as to "geological" or "botanical" or "historical" geography. It is indeed true that the production which we observe or record can take place only if both physical and economic conditions permit. But if the ultimate task of the geographer is interpretation, his basic task is description. We answer the question "Where?" fully and accurately before we attempt the further question "Why there?" And our answer to the latter question must take into account both physical and economic factors as well as others which fall into neither category. We are confronted again by the double meaning of "economic." The production of food is evidently a contribution to material welfare and is economic in that sense, but only where scarcity relationships arise to an important degree do we encounter that subtler and more fundamental meaning of the word "economic" which certainly has its relevance to geography though not equally at all times and places.

It will not be out of place here to point out that discussion of the technical processes of production has no more fitting place in geography than in economics. Some background of technological knowledge is doubtless required in handling problems in both subjects, but in both it is liable to obscure the aim. In geography the root of the trouble lies in the entirely defensible practice of teaching at the school stage a good deal of what is really elementary sociology and "general knowledge" under the heading of "Geography." But it cannot be too strictly insisted that to study the life and work of the inhabitants is not to study the geography of a region; it is merely to acquire certain ancillary knowledge to further geographical interpretation.

Similarly, the knowledge of the processes that take place inside a blast furnace in no sense confers an "economic" aura on geographical work; it merely supports such work with a little relevant technology. It is desirable knowledge, but economist and geographer alike can in the final analysis take manufacturing processes as given: it is enough that a suitable mixture of ore, limestone and fuel yields the required product.

We reach then a second stage of our conclusion. Our study does not become in any real sense "economic geography" because on occasion we are forced to reckon with a purely economic factor. Still less does it merit this title because we are perforce concerned with economic activity in the course of regional description and interpretation. We are then concerned simply with geography in the fittest sense of the term—not with a sub-division of it.

We can now clearly appreciate the distinction between the actual and possible character of economic geography. On the one hand, there is a large body of geographical works, variously entitled "Economic Geography" and "Commercial Geography," which serve a useful purpose, above all for reference, but which appear on inspection largely to contain the familiar matter usually conceived as "pure" or "general" geography. Such books appear in successive editions and clearly meet a demand: in this sense at least they are economic. In other respects their "economic" character is unsatisfying, nor can they claim in any deep sense to be "link" studies between economics and geography. The reader who sought to learn, for example, where cotton is grown and the trade relations in this commodity would doubtless find satisfaction. If he enquired further about why cotton is grown where it is, he could discover there certain physical considerations of relevance (e.g., climatic factors) but not a wholly satisfactory answer.

Certainly he would be unlikely to find in the economic geographies a new and inviting trail blazed through the borderland of geography and economics, a presentment of the continual interplay of economic and physical factors in a world where equilibrium exists only when it is assumed. What he finds, in short, is of a modest and useful character, an economic geography, to use Sir H. J. Mackinder's words "concerned with the raising and distribution of commodities." This the student will readily recognize as "regional" geography with special reference to agriculture, industry and trade. He will require for the study of such texts little acquaintance with the science of economics; nor indeed will their study lead him even gently into the mysteries of that social science.

Standing distinct from the text-books and pointing the way to an economic geography which could be fittingly conceived as a specialist branch of geography are research contributions of substantial interest. We would refer to the mode of approach adopted by the author of *The Pastoral Industries of New Zealand*. In the preface to this work Dr. R. O. Buchanan described his conception of Economic Geography as follows:

> "Put briefly, it is that, in industries organized on a commercial, as distinct from a purely subsistence, basis, geographical conditions express themselves, if at all, in economic, mainly monetary, terms; and that the nature and extent of the influence of the geographical conditions are themselves dependent on the precise nature of the economic conditions. That will be recognized as merely specifying one type of instance of the generally accepted view that geographical values depend on the cultural stage achieved by the human actors. If that argument be accepted, it will follow that the geography of production must be a study of the *interaction* of geographical and economic conditions in the area and for the products concerned, and it is such a study of mutual cause and effect that is here attempted."

It is clear that the objective thus set involves fairly and squarely a task within the field fittingly described as economic geography, for the author sees in the production of New Zealand and its related geography the dynamic expression of interacting physical and economic conditions, the latter both local and world-wide in character. He is concerned in his study not merely with considerations such as climate, soils, drainage and accessibility, but also with world prices (for wool, meat, and dairy products) and with the quotas, tariffs and prohibitions of the populous industrial countries of Europe which react on the price level of the world market and thus on the agricultural geography of New Zealand. Similarly, as another illustration of how geography and economics can interlock, reference may be made to articles by Dr. W. R. Mead which examine, with reference to Finland, a specific application to agriculture of the concepts of both the physical and the economic margin.

CHAPTER VI

POLITICAL GEOGRAPHY

"Don't start that hare! God knows where it might lead us!"
Cited by Y. M. GOBLET, *The Twilight of Treaties* (1936).

WITH these words a European delegate at the International Geography Congress at Paris in 1931 reacted to the proposal that a section dealing with political geography should be created to function at subsequent meetings. Doubtless his fear of the possible corruption of academic geography by such an innovation is easily explicable in terms of the circumstances of the time. The first world war had undermined the European states system and, in its effects, recast the political map. Old empires had dissolved; new national states had emerged. As Count Keyserling decribed it, Europe had become "an astoundingly manifold, astoundingly riven structure; the Balkans constitute its truest prototype."

The small area of Europe, equivalent to one-twelfth of the earth's land surface, had come to contain uneasily within its bounds three-fifths of the independent states of the world, and these were enclosed within no less than 17,000 miles of boundary (the comparable figure in 1914 was 13,000), or one-sixth of the world total. Moreover, it was then a fresh recollection in the minds of geographers who had taken part at, or watched the activities of, the Paris Peace Conference that not only had geography played there a sober and scientific part in the examination of boundary problems but had been misused by those whose prime concern was the advancement of particular territorial aspirations. The maps produced at the Paris Conference, though for the most part the result of careful expert study, included others the authenticity of which was at the least suspect.

> "A new instrument was discovered," wrote Dr. Isaiah Bowman,[1] the doyen of American political geographers—"the map language. A map was as good as a brilliant poster and just being a map made it respectable, authentic. A perverted map was a life-belt to many a floundering argument."

And, spanning the fields of geography and politics, the new "science" of *Geopolitik* (geopolitics) had been born and was already establishing itself firmly in Germany.

The natural apprehensions of the geography delegate could not, however, obscure the fact that, as a major war inevitably emphasizes, problems of a politico-geographical character abound and, further, that a subject of formal study, known as political geography, had long been studied in Germany and had behind it a philosophic ancestry in the literature of ancient Greece. When Aristotle, in his *Politics*, discussed realistically the territorial basis of the ideal city state and the site requirements of its capital and strongholds, and speculated on the relationship between climate and political organization and international relations, he foreshadowed a line of enquiry which, as Political Geography, took scientific shape in the writings of Friedrich Ratzel in the late 19th century. This field of enquiry concerns the relationship between geographical and political phenomena and arises out of the fact that states, as politically organized communities, have inevitably a territorial basis and a geographical location. The nature of this relationship was for a long time rather guessed at than understood, and broad generalizations were made, such as that of Montesquieu, that forms of government and political institutions were the result of differential climatic conditions. And it was only with the advent of scientific geography (see Chapter I) that close study of particular states and of particular political problems could be fruitfully undertaken and a useful geographical contribution offered.

Our present task is to indicate the nature and scope of political geography and its place within geographical studies as

[1] Cited in *What Really Happened at Paris*, edited by E. M. House and C. Seymour (1921).

a whole. We shall be concerned less with precise definition, which can serve only to confine a growing study, more with the problems it attempts to elucidate and the method which it employs.

At the outset and to avoid confusion we must examine the concept "geopolitics." The term derives from Rudolf Kjellen, Professor of Government at the University of Göteborg, whose book *The State as a Form of Life*, published in 1917, had a great success in Germany, where his doctrine of the State as an organism which could grow territorially, if necessary by resort to war, fell on fertile soil at a time of national defeat and weakness.

Under the able direction of General Haushofer, head of the German Academy and editor of the *Zeitschrift für Geopolitik*, geopolitics were intensively developed in Germany to provide a geographical and intelligence background to Nazi political aggression and to spread the idea of global thinking in a world where great prizes of territory and political power were to be won by Germans if they had the will, direction and knowledge to attain them. Ironically, Haushofer derived his basic idea from the British geographer Sir Halford Mackinder whose generalizations and prediction about the "Heartland" of the "World-Island" led Haushofer to believe that Germany, if associated with the Soviet Union, Japan and, if possible China also, could exploit the possibilities of this vast continental base with its interior lines of communication and achieve world mastery. In short, in German dress, geopolitics became, as A. Demangeon put it, "a national enterprise of propaganda and instruction." And in view of the so-called principles of German geopolitics, which had no scientific validity but were designed to give support to a policy of national aggrandizement, Dr. Isaiah Bowman summed up *Geopolitik* as "illusion, mummery, an apology for theft." Even so, properly conceived and soberly studied, geopolitics may fittingly be regarded as an extension or application of political geography to the *external* geographical relationships of States.

Political geography focuses attention on both the external and the internal relationships of states, each of which raise problems. This study, it should be further explained, invites two lines of approach. There is first the effect on the present-day geography of political action; secondly, the significance of the geography behind political situations, problems and activities.

In the first case, the political are one of the many different human considerations with which the geographer studying the world or its parts must be concerned. Such considerations must be appreciated as part of the general subject matter of geography, although they may become matters of specialist study by the political geographer. As a result of historical processes, some of which have been operative in very recent times, a world political map has been created, defining patterns and distributions, related to the physical divisions of the earth's surface, the importance of which in human geography can hardly be overestimated. States demarcate, in fact, what are perhaps the most significant of all the many cultural regions that the geographer distinguishes, regions, too, contained within limits more precisely defined as a rule than those of other cultural regions and even of physical divisions.

The regional geographer cannot ignore the political and administrative divisions of the area with which he is concerned; indeed, regional geography must concern itself *inter alia* with political considerations and also political units, thus recognizing their claim to "regional" character. A regional study of Germany, for example, although it would focus attention primarily on its distinctive regions as indicated by structure, physique and economic and social phenomena, would have to recognize also as regionally significant the administrative zones of the occupying powers as well as the provisional boundaries within which Germany has now contracted (Fig. 7). Similarly, the regional geographer of the Soviet Union would take note of the sixteen constituent republics which form political regions, although they may bear only loose relationships with the physical divisions and other human regions which he discerns.

In various other ways, matters fittingly called "political", intervene to modify "regional" geography. Differences of railway gauge in Europe and Asia, due in part to political precautions, impose obstacles to transport. Transportation by land, air and inland waterways does not necessarily follow the obvious geographical lines regardless of political frontiers, even in peace-time and notably in Europe. Air routes, for example, depend for their alignment on the permission of, and the facilities provided by, particular states in respect of the land and air space of state territories. Contrast the rivers Volga and Danube, the one wholly contained within one state, the other flowing within seven states, and the consequent effects of this

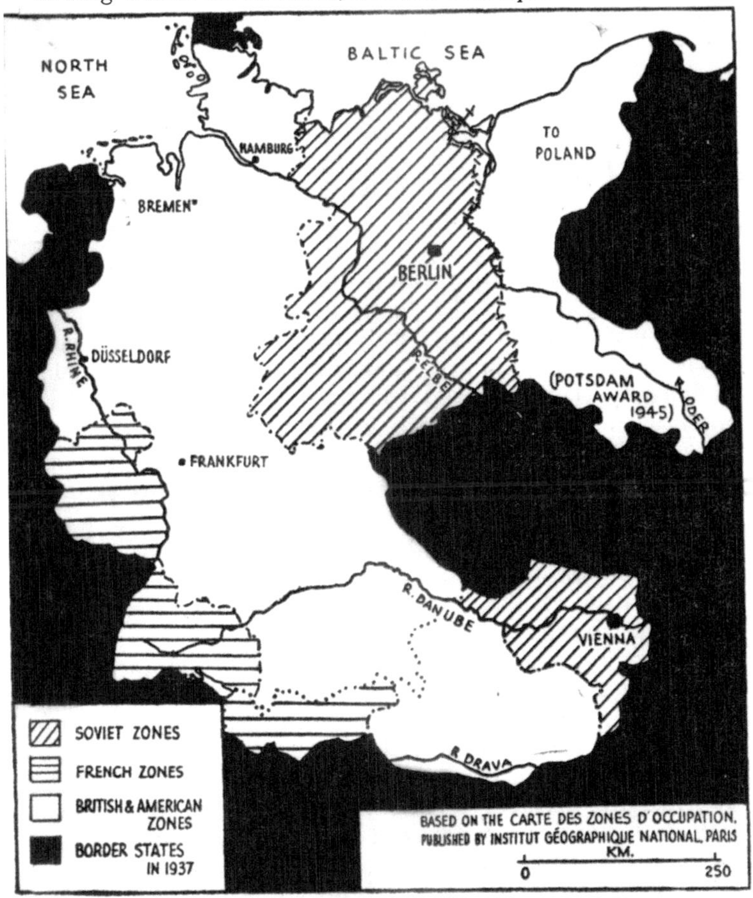

Fig. 7. The Partition of Germany and Austria: the Remnants of *Mitteleuropa*. Note that Germany has been both partitioned and truncated.

at the present time on their potential development for navigation, irrigation and hydro-electricity. The flow of trade may be directed along this or that channel as a result of differential rates, customs dues and other such means which are political in character. The migration of peoples is politically controlled—witness, as a remarkable instance, albeit a grandiose gesture rather than a prepared and practicable plan, the decision of Argentina to receive no fewer than four million immigrants from Europe, starting at the rate of 30,000 a month.

The economy of modern states becomes increasingly, and in the U.S.S.R. is wholly, a function of the state, the decisions of which produce geographical changes—in the location of industry, in the extent and productivity of agriculture, in the network of railways, in the creation of new towns and in the growth of existing ones. In a word, the geographer, although he may not be concerned with political geography specifically, must be cognizant of the political factor in geography, the importance of which varies from place to place. This factor tends to dominate the geographical background, for example, of Poland, Manchuria and Palestine; in North America, in contrast, it appears recessive, although even there politics, inter-state rather than international, enter into projects and activities of geographical interest. Thus since, it is argued, the best unit for an integrated water-planning scheme is the river basin as a whole and river basins are commonly divided between states, there is clearly a political background to projects such as those relating to the St. Lawrence and Tennessee rivers.

The political geographer, as a worker with the broad field of human geography, delimits his enquiry to the common borderland of geography and politics in the belief that by applying his own methods and technique he can deepen the understanding of his own science and perhaps, too, contribute something to sciences cognate to his own. His *locus standi* rests on the fact that states have a territorial area and a space location, both of which have a certain degree of uniqueness. His subject matter is the politically organized earth, and is thus virtually coincident with the whole earth's surface, for little land remains that has not been politically allocated, and even the high seas, though not politically organized, are fields of state interest. He is however concerned with the world in only one major aspect. It is not for him to duplicate the work of other geographical specialisms, although he will note the bearing of their studies on his own problems.

Unlike the regional geographer, who attempts a balanced appreciation of a region in its related physical and human aspects, the political geographer attempts with reference to a political region or politico-geographical problem, by interpreting maps and working in the library and in the field, to analyse, assess and map his significantly different data. Whilst he must be a geographer in outlook and training, he must needs be conversant, so far as is profitable to him, with the politics and history (economic and political) of the countries with which he is concerned.

The political geographer at once encounters the fact that, as a result of exploration, almost the whole of the earth's surface is now patterned politically, and that even those marginal lands where excessive cold restricts occupancy and settlement are either held or claimed by existing states. Political territories are susceptible to classification on various lines. On ground of legal status states are distinguished as either sovereign or independent—the characteristic type in Europe and America, or dependent or colonial, as are many of the territories of Africa, Central America, and numerous islands or island groups in the Oceans.

In respect of their internal political structure, too, states are classified as "unitary," as are the United Kingdom, New Zealand and France, or as "federal," as for example, Switzerland, the U.S.S.R., Yugoslavia, the United States, Canada, Australia and Malaya. The distinction currently used in international history and politics between the "great" and lesser "powers," a distinction which rests on the military, diplomatic and economic power and prestige enjoyed by the former, should also be noted. The political geographer makes himself aware of such distinctions and terminology before he visualizes the political scene from his own point of view. When he does this, his curiosity is aroused at every turn. Why are the political patterns so various in scale and so irregularly distributed between continent and continent and between higher and lower latitudes? Why has Europe become so comparted politically? Why has Africa become so largely a land of dependent states? Is the Indian Ocean to remain, now that India, Pakistan and Ceylon have become independent and now that the United States commands the greatest naval power, what it was formerly—a "British lake"?

To such broad introductory questions he may add others bearing on the remarkable inequality of states in so many respects: in territorial extent, population numbers and hence "man power," natural resources, geographical position, transport facilities, the ethnic, national and linguistic composition of their inhabitants, levels of material culture and stages of development. Certainly there are matters here which invite investigation since they present an apparent complexity and irrationality comparable with those of the geological map and are no less significant to the comprehension of our present environment.

Resort to historical geography familiarizes us with the fact that the political map has been shaped and changed continually in response to the interplay of human forces (largely belligerent) and statesmanship in historical, including recent and contemporary, times. It is a far cry from the simplicity of the political map of Europe during the first four centuries A.D., when the Roman Empire formed the only organized state in Europe, to the changing maps of Europe during medieval and modern times, notably in our own times, as a result of the first and second world wars. Similarly colonial and diplomatic history provide the key to the emergence of the independent states of South America and to many surviving dependent territories held notably by Britain, France, Spain, Portugal and the Netherlands.

Present-day states, like the land-forms which modify the surface of their territories, though varying in age, stability and scale, have each their own story of origin and growth, some elements of which are common to them all. In their beginnings states have much to do with geography. First people occupy and settle scattered parcels of land, suitable to their needs, these lands being insulated one from the other by waste, unsettled frontier zones. Small social groups thus subsist at first in relative isolation, but subsequently relations, both peaceful and warlike, are forged between these similar and neighbouring groups. Such relationships depend on the discovery or creation of routes, for example, by way of open hill-tops or of river valleys. As time goes on, with the progress of settlement and cultivation, neighbouring groups coalesce, the waste frontiers tend to disappear, and larger social groups, commanding wider and more complex territories, are formed appendant to a capital city. Capital cities and route-ways thus appear as essential means by which state organization and functions can be achieved.

In the same way, we can discern what is called the "nuclear area" of emergent states—a region which comes to be endowed with a superior status and with superior potentialities and which, lying around the capital, contains the major endowment of the state in respect of population, food and other resources, means of accessibility and political direction. Thus in France, as it began to shape under the Capetian kings, part of the Paris Basin, astride the route Paris–Étampes–Orleans, formed the nuclear region; so also in England, immediately before and after the Norman conquest, the London Basin, containing the capital at Westminster adjacent to the city of London, played a similar rôle. Similarly, behind the 19th-century achievement of German and Italian political unity, nuclear regions which took the lead were respectively Prussia—a composite entity geographically of which the Brandenburg electorate was the core, and the duchy of Savoy, marginal to the western Alps and the North Italian Plain.

In the growth of the Swiss Confederation, one of the smallest yet stablest states of Europe, the nuclear region was provided by the tripartite union of the cantons of Uri, Schwyz and Unterwalden, grouped around lake Lucerne and extending alike into the Alps and the Swiss plateau, which took a firm stand against alien domination and gathered around themselves neighbouring peoples willing to co-operate and to combine forces within a single confederated state.

Many of the states of today originated in the way broadly generalized above. In the course of centuries of consolidation and expansion effected by diplomacy and by war, such states extended their organization and control up to the limit of their frontiers, only some of which appear to have achieved real stability. Nationality, and above all the consciousness of nationality, evoked under external pressure, were the potent forces behind the shaping of states in recent centuries, notably in Europe, but we must recognize also a type of state the origin of which is less the outcome of forces operative within its own borders than the imposition of an outside alien conqueror or acquisitor. This applied to such states as the former Austro-Hungarian and Ottoman empires which survived until the war of 1914–18, and to British India, where the political unity or association of peoples divided in race, religion and culture, was established by the British intruders.

The U.S.S.R., as heir to the Russian Empire, has by means of federal machinery created a state which incorporates numerous different ethnic and national elements, many of which were originally colonial dependents of Tsarist Russia. Clearly such states present greater complexity than do those nation states, such as Sweden and the Netherlands, where one nation has become politically organized as a state, and has grown *in situ* with its roots in the soil. This is not to ignore the fact that a degree of compulsion has entered into the formation of most, if not all, states, whether to maintain internal cohesion (e.g., the United States during the Civil War of 1861–5) or to secure desired frontiers (e.g., France under Louis XIV).

If we examine more closely the geographical content of states, it is clear that they are compounded in varying degrees of contrasted elements. State territories are not necessarily continuous: note the detachment of Ulster from Great Britain, Alaska from the United States and, formerly, East Prussia from Germany. Contrast, as extreme examples of territorial continuity and territorial diffusion, the Soviet Union and the British Commonwealth and Empire, the one dependent on long rail hauls, the other on extended seaways, and both on airways, to effect the inter-relations of their constituent parts. Again, it is generally true that state territories show no conformity with the physical regions which the geographer distinguishes on the bases of climate, structure or land-forms. It is well for states that this is so.

In so far as a state includes a number of distinct physical regions or parts thereof, it is the better endowed, since each may afford different products and the whole may contain economically complementary parts. On the other hand, it may be argued that it is both unnecessary and uneconomic for states to seek self-sufficiency, a policy born of war psychology, since nothing can be easier and more mutually profitable than the exchange of commodities by trade. But in so far as such ideas persist, it remains politically desirable to possess as the basis of state economy areas physically contrasted and complementary in their productivity. Compare, on the one hand, France, in which a varied agriculture, based on a wide range of soils and climate, is associated with industrial resources such as coal, water power, iron and bauxite, and on the other, a small state such as Hungary where roughly uniform physical conditions set drastic limits to its domestic economy. In this context it is relevant, too, to take account of the productive resources of the colonial appendages of states.

If, further, states are studied from the standpoint of their nationalities and languages, they are found to contain varied elements, the geographical significance of which (witness that of the Sudeten Germans in Czechoslovakia before the war of 1939–45) may in itself be politically significant. We need not stop to consider the ethnic content of states: it seems certain that no state possesses, though some may claim, racial homogeneity, for distinct racial groups, even if such existed in prehistory, exist today only as arbitrary differentiations of the anthropologist or as treasured figments of the imagination.

Nations, however, are recognizable and vital human groups, and in relatively recent centuries, though not throughout history, many have claimed and secured separate statehood. We cannot here discuss at length what constitutes a nation: like the concepts of space and electricity in physical science, it resists exact definition. But it is an entity more fundamental than the State which, as the war of 1914–18 showed abundantly, can be made or dissolved. A social group, bound together by common traditions and culture which are (or were) native to a specific land area, a nation may or may not be acutely conscious of its own individuality. Not all nations desire separate statehood: some, like the Scots and the Welsh, whilst nationally self-conscious, are content to combine in a wider state group.

Where nations have sought to become states and where their claims have received international sanction, difficulties have arisen in defining the limits of their territories. Poland, for example, has assumed many different territorial shapes during the centuries; witness its extent today and that of the inter-war period 1919–39 (Fig. 8). Not the least of the difficulties has been that, although nations tend to occupy particular compact areas, they have also often scattered "outliers" and contain alien "inliers," since "in international affairs birds of a

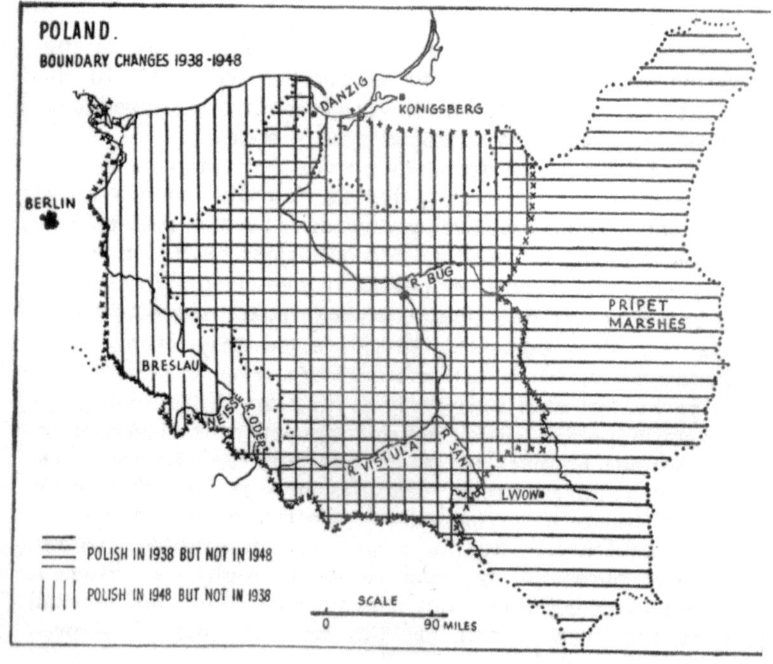

Fig. 8.

feather do not all flock together." Contrast western and Scandinavian Europe, where national groups are compactly and simply disposed, with east-central, including south-eastern Europe, which contains many areas of mixed and scattered nationalities: Bessarabia, now re-united to U.S.S.R.; upper Silesia, no longer split between Germany and Poland but allocated to the latter; Macedonia and the Banat of Temeşvar. In some areas, too, e.g., western Transylvania (part of Romania), the urban population is of one nationality (mainly Magyar) and the rural of another (mainly Romanian). Only exceptionally have drastic attempts been made to solve these difficulties by the removal and/or exchange of populations.

This method was applied in the 1920's by Greece and Turkey, and recently such large-scale transfers of population have been effected of Germans from Polish-occupied Germany and from Czechoslovakia, and of Poles into Poland from Soviet-occupied eastern Poland. Certainly, minority problems so-called characterized the inter-war years 1919–39 and afforded scope for mischief makers and aggressors. It is therefore important to note the bearing of this on the internal problems of states. In so far as a state includes several nationalities each seeking specific and disharmonious ends, so far its strength and cohesion are weakened, the more so when, as in Yugoslavia, the separate groups stand at different levels of material culture and occupy physically separate compartments of the country.

Further, although language is commonly the distinguishing symbol of nations, the frontiers of language and nationality are seldom coincident. Not infrequently a nation has more than one language: the Swiss have four; the Canadians and Belgians have two. The variety of tongues may be reflected in divergent policies of competing groups, as in Belgium and the Union of South Africa, although much less markedly in Switzerland and Canada. Several nations possess the same language; thus English, Portuguese, Spanish and French are languages of extra-European nations. Similarly, many existing languages owe parentage to a common source, witness the many which have a Germanic ancestry, but this carries no suggestion of national and political affiliations. The language patterns, so many and various in Europe, have a relative simplicity in the western hemisphere. There are, for example, only two official languages—Spanish and Portuguese—in virtually the whole of South America.

The state has external no less than internal relations, and forms in this respect a unit in international law and relations. Here the political geographer is concerned with frontiers and boundaries, which are distinct conceptions: the frontier is a zone, the boundary a line. Since almost the whole of the land surface of the earth has been politically comparted, all states, except some insular ones, have frontiers where their territories adjoin. Not all states have boundaries, however, since although these have usually been *delimited* in international agreements (treaties, conventions, etc.), they are not invariably *demarcated* on the ground. The political frontier, whatever else it may be, is certainly a geographical expression, and may, indeed, be conceived of as one kind of geographical region.

Geographers are familiar with many kinds of frontier, apart from the political. But whether they are concerned with frontiers of climate, of vegetation, of settlement, of language or of states, they are thinking in terms of zones rather than of lines. Ratzel long ago aptly illustrated the true nature of the frontier when he instanced the tidal delta as a pure frontier in physical geography: standing between land and sea and alternately part of each, the delta is essentially a region of transition, a zone of differentiation. In a similar way, the political frontier, lying at the junction of distinctly organized political groups, is characterized by transitional features, more or less marked.

Many attempts have been made to classify frontiers. We may set aside that which distinguished "artificial" and "natural" frontiers: all frontiers, since they are zones, are natural in that they have inevitably a physical basis; so, too, in varying degrees they are artificial in so far as they have been transformed by human agency. Again, it has been suggested that frontiers are either "dead" or "living"—dead where equilibrium between the contiguous states has long been established, and living where they are still subject to inter-state pressure and thus liable to change. The geographical study of frontiers, like that of vegetation, has been not a little confused by the notion of the "natural" frontier. This term has been widely used to express at least three different ideas. First, desirable or ideal frontiers indicated by bold physical features: thus for France the Rhine, the Alps and the Pyrenees (the old frontiers of Gaul) were held (by Frenchmen) to be its natural frontiers. Second, frontiers, whether ideal or not, which coincided with prominent physical features—rivers, coasts, mountain crests, watersheds, etc. Third, frontiers have been called natural because they present formidable barriers: in this sense of the term, the Himalayas are called the natural frontier of India.

It is clear that the interplay of political forces, not any inert control of nature, governs the formation and shifts of frontiers, and in this sense it is true to assert that "there exist in nature only the frontiers which we select." But according to the kind of frontier desired or established, different kinds of country have differential values. In fact, two distinct types of frontier may be sought and two distinct types are found, namely, frontiers of contact and frontiers of separation. Hitherto, almost invariably, states have sought frontiers which foster separation from, rather than intercourse with, their neighbours.

It is not necessary to look far afield to find illustrations of these types. The Franco-Belgian frontier between the western end of the Ardennes plateau and the English Channel is a frontier of contact and intercourse. There, except for a tract of low land in the west which the French can flood at will, the boundary crosses a low, well-settled plateau which offers no physical obstacles to passage; nor is it a linguistic divide, since the same language—French and Walloon, a dialect of French—are spoken on both sides. In contrast, the Pyrenean and Alpine frontiers of France belong to the other type: consisting as they do of wild, mountainous, little settled country, with transverse route-ways restricted to the depressions and passes, these are frontiers of separation.

It has been argued that the frontiers of these two types are conditioned by certain kinds of country distinguished on a physical (including climatic) basis, but this generalization, applicable though it is to historical times, has lost much of its force with the advent of the steamship, the automobile and above all the aeroplane. Mountains, forests, deserts, moorlands and marshes, because they impeded passage and could not sustain large populations, were commonly, together with lands deliberately laid waste, frontiers of separation throughout history. Open plains, low hills, plateaus and river valleys, since they afforded better accessibility and conditions for settlement, tended to provide frontiers of contact and intercourse. But no one could now claim for the desert belt of the Old World or for the Atlantic Ocean the remarkable degree of separation and protection which characterized them as frontiers of the Roman Empire; nor do the maritime frontiers of Britain insulate her to the extent which they did in ancient and medieval times. Further, the policies of states can do much to counteract geographical forces which tend to bind or to separate contiguous states. Thus, for example, trade and other contacts across naturally open frontiers in Europe have often been curtailed by political action, while by military works something of a barrier can be erected.

An historian has argued that the significant factor in frontiers is the pressure behind them and that "the more populated a district the greater the pressure." But it cannot be inferred that international friction is restricted to frontiers subject to population pressure, such as those in Europe. Witness the Gran Chaco frontier between Bolivia and Paraguay, where a virtually unsettled jungle, which could have effectively insulated both states, was made a battlefield.

The boundaries of states, in so far as they are defined in treaties and marked on maps and on the ground, are, like the frontiers which contain them, matters of geographical study. The practice of demarcating boundaries is very recent and by no means universal. The European states which were overthrown during the Revolutionary and Napoleonic wars, almost without exception, had no precise linear limits, and the boundaries of France itself were not accurately known in the 18th century. The reasons were two-fold: first, the confused entanglement of political rights to land which varied from full sovereignty to mere overlordship; and second, the fact that the arts of surveying and cartography were not then sufficiently developed.

Before the full development of geodesy and cartography,

international tension and even wars were caused because statesmen, lacking adequate maps and geographical knowledge too, defined boundaries in loose or ambiguous terms. There was, for example, a dispute at the Conference of Paris in 1856, relevant to the Russo-Turkish boundary delimitation, because there were two possible locations for a place called Bolgrad. The Treaty of Paris of 1783 delimited the Canada–United States boundary in relation to lat. 45° N. and long. 141° E. It was later discovered that the surveys by which these lines had been determined were inaccurate, and the resultant difficulties were not overcome until 1843 when Great Britain and the United States agreed to accept as valid the wrongly surveyed positions.

Three kinds of boundaries can be distinguished: orographic boundaries, astronomical lines and the limits of reference. The first are drawn in relation to physical features which are in varying degrees prominent: rivers, mountain crests, hills, etc. The other two types are determined respectively by lines of latitude and longitude, and by geometric methods, for example, by lines drawn between surveyed points. The determination of orographic boundaries depends upon the existence of large-scale maps or upon the preparation of local surveys. This kind of boundary is most commonly adopted in densely settled and well-mapped countries such as Europe. The treaty makers of Versailles in 1919 were well supplied with maps and, since in Europe every inch of the ground was contested, resort was had to maps on as large a scale as twenty-five inches to the mile. In contrast, in those parts of the world which are in process of settlement, maps on the scale of a quarter of an inch to the mile are usually adequate. Moreover, in such countries, astronomical lines and boundaries of reference, if only as a *pis aller*, are usually preferred. Thus boundaries of these types are common in Australia, Africa and the Americas. Further, such boundaries are not as a rule demarcated unless, with the advance of a settlement and economic exploitation, the frontier regions acquire economic or strategical value.

Occasionally an astronomic boundary has been marked out on the ground and recorded on large-scale maps in land which was unsettled, as in German South-West Africa, where the work of demarcation in the Kalahari desert involved loss of life and considerable cost. The most remarkable instance of an astronomical boundary which has been marked out on the ground is that between Canada and the United States which was completed only in this century. It extends for 1,300 miles along lat. 49° N., westwards from the lake of the Woods, to the Gulf of Georgia. Before the exact definition of this boundary many disputes arose between the two countries, yet this highly artificial divide, lying within what is essentially a frontier of contact, is the most striking example today of a frontier of equilibrium. It points the obvious moral that frontier tension is a function not of nature but of man.

We have attempted to outline in general terms the nature of political geography as one of the specialist branches of geography. Clearly it involves study of many topics additional to those to which we have alluded here—the contribution of geography to state-designed plans, the nature and relationships of capital and provincial cities, the differential growth or decline of the populations of states—and hence of their economic and military man power. We may conclude by suggesting that a geopolitical approach can provide an illuminating background to many complicated international problems.

A characteristic feature of the inter-war period 1919–39, as indeed of the present time, is the continual and sporadic disputes and tensions which engage the interests and attentions of the Great Powers and of international bodies, formerly the League of Nations and now the United Nations Organization. The understanding of such problems in those many countries where democratic institutions exist and where therefore the citizen should attempt to form a view, is clearly very difficult and requires that background of realistic, systematized data which the political geographer can provide. "Manchuria," "Abyssinia," "Czechoslovakia," "Danzig," and now "Palestine," "Hyderabad," "Kashmir," "Batavia" and "Korea"—these and similar territorial concepts recall international problems, the magnitude of which stood (or stands) in inverse proportion to popular knowledge of their geographical backgrounds.

Nor is it merely that distance adds obscurity to the view

when a British Prime Minister (Mr. Neville Chamberlain) could describe Czechoslovakia as a distant country about which we knew little, and a French geographer (J. Ancel) could refer to Europe as the true *terra incognita* of our anxious and difficult times. We cannot here illustrate in detail our claim that, by relating international problems to their local and spatial settings, political geography can contribute substantially to their comprehension. Fortunately there is increasing recognition that a knowledge of geography can often be useful in statesmanship no less than in the art of war.

CHAPTER VIII

REGIONAL GEOGRAPHY AND THE THEORY OF REGIONS

For the land, whither thou goest in to possess it, *is* not as the land of Egypt, . . . where thou sowedst thy seed, and wateredst *it* with thy foot, as a garden of herbs: But the land, whither ye go to possess it, *is* a land of hills and valleys, *and* drinketh water of the rain of heaven. . . .

DEUTERONOMY, xi.

Nature and Art in this together suit:
What is most grand is always most minute.

WILLIAM BLAKE.

IN the last five chapters we have reviewed fields of knowledge, each of which is evidently a branch of geography, contributing its part to the geographer's full equipment and also affording scope for specialization. Each of them makes contact with one or more other science and could in no sense be studied in complete ignorance of these sciences.

Cartography and Historical Geography stand rather apart from the rest. The former is essentially a subject in its own right and not a branch of geography, yet the dependence of the geographer on maps is so great that the subject or large parts of it becomes his by adoption and use. Historical Geography can be viewed from two standpoints. In so far as we restrict its proper aim to the reconstruction of past geographies, in the attempt to "reclothe" the historic present for selected past times, it is necessarily a specialized field demanding the full rigours of historical scholarship. In a more general sense it is true to say that the point of view and methods of thought of Historical Geography must inform the whole subject, even if our main aim is concentrated upon present or recent times. It is necessary to see the geography of any area as, in effect, the resultant of a series of superimposed geographies, to be aware of the historical antecedents of each and every pattern of the landscape, both of existing features which have developed and are developing, and of anachronisms surviving from former patterns and processes.

Physical Geography, and to use a convenient general title, Human Geography and its sub-divisions (economic, political, demographic, etc.), also rest upon background disciplines, but their formal rôle in the subject as a whole is rather different. We need not here re-argue the basic importance of physical geography but, as we can now appreciate, it is no more basic than social geography; neither really exists without the other. The important point is that in these branches and their sub-divisions a considerable element of generality exists, or stating the matter alternatively, they can be pursued as systematic studies. They are in fact branches of what has been termed General or Systematic Geography as contrasted with Regional Geography.

We have seen that this distinction was drawn in the 17th century by Varenius; it was discussed at length during the 19th century, especially by the German geographers. Sometimes it has appeared that there is an antithesis between these two main aspects of the subject, and one or the other has been preferred as the more hopeful line of advance. In later years there has been wide concurrence in the view that regional geography is at least a vital and indispensable part of the subject; some have not denied it the status of "the culminating branch." In what sense and to what extent this is true we must now enquire, for the serious student of geography will find much of his time occupied in its study and for the general reader it is and always has been geography *par excellence*.

This question has recently been discussed carefully and at great length by Professor Hartshorne who, following Hettner, represents the systematic sciences as lying in a plane of their own intersecting that of geography and each with its geographical equivalent or derivative. The detailed "allocations" in the geographical plane are indeed open to not a little criticism. With meteorology goes climatology—which is well enough, but geology contributes to the geographical plane only "Land Forms." This no doubt represents an undoubted latter-day prejudice of some geographers against geomorphology based upon the belief, entirely logical but doubtfully psychological, that it is really part of geology. Letting this formal point pass, however, we note that geology contributes much more to systematic geography than "land-forms," for rock characters, soil types, mineral distributions and surface and sub-surface hydrology are by no means to be subsumed under this head. Botany and zoology present no difficulty with their projections as plant and animal geography, but directly we enter the human or social field, there is scope once more for at least amiable altercation.

We have discussed the relation—or perchance lack of relation—of economics and economic geography, but the faults there are such as time may cure. Political Geography may be permitted to link with Political Science, although the latter is both ambitious and ambiguous. Hartshorne also recognizes a Social Geography co-ordinate with Sociology. Perhaps there ought to be such a subject and conceivably one may grow, but it is difficult to see its field if economic and political geography each claims its own. The present writers see no reasons for preferring Fitzgerald's suggestion of "social geography" as an acceptable general name for what has often been called human geography and includes economic and political geography among its sub-divisions. Within this group too both physical and social anthropology have their geographical correlatives, for which logically the term anthropo-geography may serve. For some it has no doubt a somewhat different and wider connotation equivalent to what we have termed social or human geography, but although this usage, deriving from the writings of Ratzel, has its historical warrant, it has little else to recommend it.

All such points of detail on matters of nomenclature will no doubt remain subject to differences of opinion; we cannot hope to propound here a universally acceptable scheme of nomenclature. A more important issue is raised in Hartshorne's limitation of the field of systematic geography and its branches to the study of "areal differentiation," to the distinguishing of *areas* whether in the world as a whole or in some large part of it. This, for instance, is equivalent to saying that we must not, for example, study land-forms for their own sake or with a view to understanding their origins, but simply in so far as they enable us to construct schemes of morphological or physiographical regions. Similarly it would follow, from his insistently emphasized dictum, that systematic economic geography is a tool for resolving the earth's surface into parts or at least for establishing differences between one part and another, not for throwing light on any aspect of descriptive or analytical economics.

In such a distinction there is a manifest element of common sense, yet literally applied it is too narrow. We may for convenience think of subjects as lying in planes of discourse but this carries no implication that we should be restricted to one plane and forget the universe of discourse. It will be evident that we have not here accepted such a restriction in the aims of such a subject as Political Geography. If it throws light on the areal differentiation of the earth well and good, but we should not wish to deny the political geographer the right or the hope of making contributions from the geographical side to political science. Restrictions such as the one suggested arise from the mistaken notion that subjects are defined by their aims rather than by their methods.

We may take an example from geology, a subject studied by many geographers who may therefore be able readily to see the analogy we seek to draw. The prime (or ultimate) aim of geology may no doubt be defined as the elucidation of "earth history." But the igneous petrologist has become increasingly preoccupied with the physical chemistry of rock magmas and their crystallized products. Although the final conclusions and interpretations based upon such work may make a large contribution to earth history, this contribution is by no means direct or immediate. Yet it is of no service to accuse the unfortunate petrologist of being no geologist and of making contributions to physical chemistry, nor is there the faintest reason to suppose that if the geologist did not do the work the physical chemist would do it for him. The limitation of geological investigations to those which make a direct and evident contribution to stratigraphical (historical) geology would ignore the past history and existing state of the science and the interests and aptitudes of those who pursue it. No less would it be a foolish and retrograde step to force geography too straitly into the Procrustean bed of "areal differentiation" in obedience to a seemingly logical definition.

When all has been said, it remains true that there is value and a contribution to clarification in Hartshorne's analysis. When he claims among the functions of systematic geography the production of generic concepts, he is surely on unassailable ground. When systematic geography recognizes "regions" of any sort, they are generic regions—types. This is evident with the several variant schemes of "major natural regions," as originally conceived by Herbertson. In reality these always prove to be climatic or climato-vegetational regions—elements of a broad repeating pattern resulting from continental and planetary situation. We need not labour the obvious case.

Less systematically but still definitely there are repeating patterns of structure and of morphological types. The high intermontane plateaus of the western Cordillera in the Americas and of Central Asia have significant elements in common. A knowledge of the English Weald assists the interpretation of the Black Hills of Dakota. Widely different as the two areas are in situation, climate and cultural landscape, they are both unroofed periclines with inward-facing cuestas; a knowledge of the one helps in the interpretation of the other, at least, in explaining the relations of drainage to relief.

It is of course clear that geomorphology abounds in such generic concepts of prime utility in regional analysis. A consideration of the fact goes far to explain why, despite much logical hair-splitting, geomorphology remains an indispensable tool to the geographer. So it is likewise in the fields of climatology and biogeography. They provide not only terminology but unifying ideas for regional analysis. The generic concepts yielded to date by social geography are less numerous. There is less tendency for even partial repetition among the cultural landscapes of the world. At a very simple level "conurbation," "town," "village," "port," "farm," etc., are no doubt generic concepts, but they hardly derive from any recognizable branch of systematic geography. But although the list is a short one, we can extend it far beyond this point. It does not need a geographer to recognize a town, but it may well take a geographer to devise a significant classification of towns or of ports. "Gap towns," "bridge-ports," "hinterlands" are generic concepts, and the same may be said of "nodality" or "nodal sites." It is, indeed, a very worth-while exercise to compile a list of such terms signifying features or relationships—we may, perhaps, even say "areal relationships" in deference to the fastidious terminology of our subject. At best the list proves all too short and simple when we have left the field of physical geography, but we are entitled to recall that the effort is still a young one.

Very little reflection will assure us, however, that the rather scanty equipment of terms and ideas we have so far devised for interpreting social phenomena "in area" have derived less from a sedulous application of the discrete branches of social geography than from what may be called "comparative regional geography." We must reaffirm our former conclusion that there is, as yet, no such subject as a generalized human geography. It is no doubt possible to maintain that such a subject is, in the nature of things, impossible. It is perhaps equally reasonable to counsel patience and hope. Detailed study cannot fail to increase the range of our generic concepts in all the fields of geography. The attack on the problem has been made by too few workers for too short a time. None the less, as Hartshorne maintains, the nature of the subject imposes inevitable limits on the possible amount of generalization. In the ugly and unfamiliar jargon of the philosophers who classify knowledge, it attains to little nomothetic quality. Geography is essentially idiographic. It is for this reason that, in the narrower sense of the term, it is denied the title of science.

It is the idiographic quality of geography that enables us to see regional geography in its right perspective. It is clear that, however successful the geographer might be in recognizing generic regions, the world shows in fact a large number of specific regions, specific in the sense that there is only one of each. Geography, like history, never repeats itself in detail;

the specific regions are unique. Prominent among such regions are the nation states, but in the larger of these there are evident divisions which we adopt as a matter of common sense or convenience. We need not in the first instance make any extravagant scientific or philosophical claim for such regions. They are "given" to immediate and unaided observation.

No one will doubt that within Britain, East Anglia, the Central Lowlands of Scotland, and the South-west Peninsula attain the status of regions in this sense, nor that in the United States, the North-eastern industrial area, the Middle West, and "the South" are, in their ways, of equally distinctive character, even though we may be in doubt about the precise position of their boundaries. In a very real sense these specific regions are more important and clearly marked than generic regions: they are the actual and recognizably distinct parts of the several land areas concerning whose differences geographers and laymen alike are legitimately curious. The necessity of the study of such regions needs no defence; it is "special geography" in the sense originally used by Varenius.

From some points of view it would have helped to avoid many confusions if we had continued to use the term "special geography." "Regional geography" seems to imply rather more. When we enquire into the nature of these implications, we find ourselves in trouble with the word "region," which is exceedingly ambiguous and heavily overworked. It has its long established and unassailable use in ordinary language; no secondary meaning which a geographer may seek to read into it can prevent our speaking and writing of the London region or the South Wales region.

In this common sense usage of the term it is specific regions that are in question, even if they may be vaguely apprehended and quite undefined. The generic region, as we have seen, is a distinct concept. But these two usages by no means exhaust the current implication of the term. Is a region to be taken as homogeneous, essentially alike in all its parts, and if so, in which one or more of a number of distinct respects? Or is it alternatively synthetic, made up of a number of contrasting though related parts—a unity in diversity?

The former sense of the word is usually implicit in physical regions. The definition of structural, morphological and soil regions assumes that there is substantial unity throughout, and the basic idea has been usefully extended into the biological field, at least in plant geography. Complete unity or uniformity cannot of course be claimed: there is always differentiation within such a region.

The matter has been very clearly treated by Bourne in a forestry survey. His essential unit is the "site," in the same sense as the term has been used in soil survey. He wrote: "a site may be defined as an area which appears for all practical purposes to provide throughout its extent similar local conditions as to climate, physiography, geology, soil . . ." While a site might be unique, more often the same type of site is to be met with again and again within some readily identifiable area.

Consider, for example, an area with rounded hill-tops intersected by valleys. Each hill-top may be so uniform in itself and so similar to other hill-tops that it may be considered as representative of a single site. The valley slopes on the other hand may be much more variable. The upper and lower slopes may be easy and the central section steep: they may all, however, differ as to soil, the upper slope having a rather shallow and stony soil and the lower slopes a deep and moist soil. Again the upper slopes facing in one direction may be exposed to winds and the middle and lower slopes facing in another direction may be sheltered from the sun: thus the upper, middle and lower slopes may have distinct local climates. Nevertheless, a particular slope facing in a particular direction may be uniform and be met with again and again and, if so, forms a "site." Finally, the valley bottom may be divisible into two or more sites largely as a result of the widening of the valley. Within the same area, each valley may be divisible into the same number of analogous sites. Differences in outcrop and soil occurring within any one of the obvious sites would necessitate its division into two or more actual sites, according to the circumstances. Bourne goes on to draw the evident conclusion that "an association of sites constitutes a distinct region," and that the crossing of a regional boundary is marked by the appearance of new types of site. In a well-marked case it would mean, indeed, the wholesale disappearance of our recurring group of sites and the sudden appearance of another, but cases of more gradual transition are obviously not precluded.

There is much of geographical value and interest in Bourne's conception. Applied in such an area as he was studying, part of the scarplands of the English lowlands, the "regions" inevitably turn out to be geological outcrops, and a conceivable criticism would be that the method constituted a lengthy and circuitous approach to the evident fact that in every sense the successive outcrops are distinct "terrains." Such criticism would be misplaced. The concept of sites and association of sites forming regions is essentially geographical, whether one thinks in terms simply of "areal differentiation" or adopts the latterly unpopular environmental approach. The object of Bourne's analysis is the scrutiny of the conditions relevant to tree growth and, although he expresses the matter more clearly, he is thinking in terms universally accepted by pedologists and plant ecologists. The concept of site is in no sense restricted to areas with a simple and marked lithological differentiation. Nor does it constitute a mere recognition of morphological areas or relations for their own sake. It implies *Zusammenhang*: the readily recognizable morphological patterns are related to local climate and soil conditions, and thus to the patterns of plant life.

If the relationships stopped here, the method would still be of use to the geographer. But in fact they go much farther. Geographers legitimately speak of the "sites of settlements," but in reality all portions of the cultural landscape are "sited"—roads, bridges, fields, etc., as well as farms, villages and towns. The regular and repeating patterns of such features convey the unity or uniformity of many a landscape, and the details are worth analysis both on the map and on the ground. Over considerable tracts we may find roads on ridge tops, farms in valleys, hanger-like woods on steep valley slopes, villages on spring lines and, as important as any of these, a definitive pattern of land utilization of which the elements reflect site characteristics.

We may clearly detect the concept of site in the important

studies of Dr. Alice Garnett on insolation and relief in mountainous regions. Not less does it inform the well-known study of V. C. Finch on the landscapes of south-western Wisconsin, and the sample studies of R. S. Platt of Latin America. Its current importance in geographical thought provides one more indication of the relevance of geomorphology, properly so-called, to geography and might well be pondered by those whose admission of this discipline to the geographical family is lukewarm and grudging.

The concept of site and of regions which are site associations provides one method of sub-dividing area. Many, though not all, of the French *pays* are regions of this type. Such regions have some slight element of generic quality: a dissected plateau is always a dissected plateau. In much higher degree, however, they are specific, depending for many of their features upon climatic environment and cultural impress. The same sites would not be "generated" even upon the same morphological basis by either an equatorial or a Mediterranean climate. Different cultures and economies will develop similar site groups differently.

Upon this fact the social geographer very properly puts a strong emphasis. Of what significance, he may be tempted to say, are the common characteristics, if any, of sandstone plateaus in Brazil and in Europe? Our reply must be that one cannot effectively isolate the differences between two regions until the ground has been cleared by recognizing their similarities. A more serious limitation of the method of division into regions which are essentially site associations appears when we pause to consider the relation between such regions when they adjoin one another. One may, for example, conveniently recognize in Belgium, Maritime and Superior Flanders, Kempenland, the plateaus of Brabant, Hesbaye and Condroz, the Pays de Herve, the Famenne depression, the High Ardennes and Belgian Lorraine. Each of these areas is a relatively compact "region" and each may be defined in simple geological or morphological terms which could, if required, be translated into the language of "sites." Yet the whole forms a mere mosaic: each region is part of Belgium, and none is self-subsisting or without relation with its neighbours.

It is here that we see emerging an entirely different concept of a region, one which shows unity in diversity. It is admirably illustrated by many old-established English parishes. Evidently the parish is, or at least was, the village region, yet its boundaries are in no sense coincident with physical boundaries (Fig. 9.)

Throughout the English lowlands, the characteristic strip parish recurs, combining several elements of scarp-crest, scarp-face and scarp-foot zones, together perhaps with an area of valley floor with its valuable water meadows. The human boundary sets itself normally to the "grain" of the country with its contrasting terrains so as to include and, in a sense, to integrate samples of the whole. The strip parish illustrates the principle only in microcosm and, in its origins, reflects conditions long since superseded. Yet the generality of the phenomenon is not in doubt. Vidal de la Blache states it "as a general principle that human establishments (i.e., settlements) preferably select lines of contact between different geological formations (i.e., outcrops)." This is perhaps too narrow and special a manner of stating the relationship.

Many well-marked lines of change in "physique" tend to show similar relations, and many, but not all such lines, mark the line of contact between different geological formations. The lines or zones of contact between mountain and plain, desert and grassland, grassland and forest tend in the same way to be zones of "geographical reaction" expressed in the peculiarities of settlement. The towns of "the Highland line" in Scotland, of the Alpine foot, and the *fontanili* line in northern Italy, of the Fall line in the United States, and many others all support this idea. It would indeed be a ludicrous travesty of the facts to claim that settlements arise only in such situations. Yet the pattern is a repeating one and has evident significance. We can see the settlement centres on such lines as the meeting places and market places of contrasted products or, in some cases, and more generally, as the contact points of contrasted cultures. The nucleated settlement, village or town, once established, tends to draw into some sort of unity parts of the contrasted countryside which serves it and which it serves; it acts as a focus for common living.

It is immediately apparent that there is a veritable hierarchy

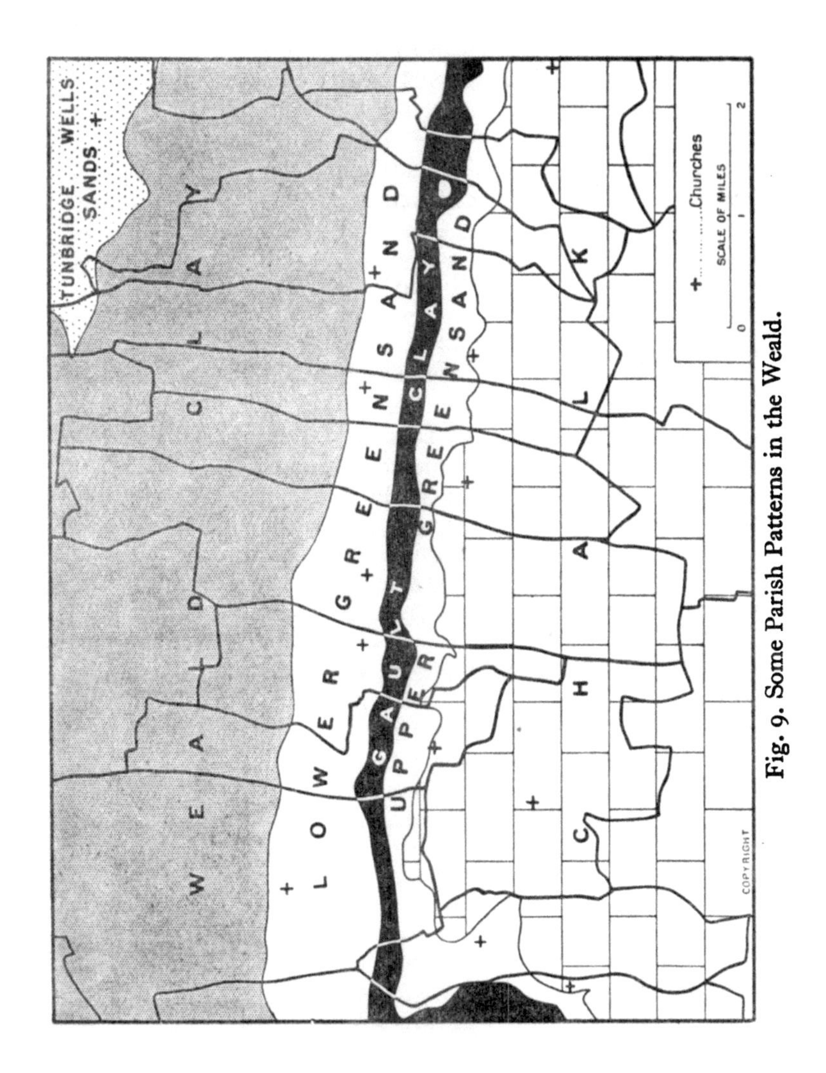

Fig. 9. Some Parish Patterns in the Weald.

of "regions" of this type, related to their urban foci large and small, and this idea has proved attractive not only to the geographer, but also to students of the other social sciences. It was early perceived by H. G. Wells in his *Anticipations* (1900), which recognized "the fundamental changes in the scale of human relationships and human enterprises brought about by increased facilities of communication," and he applied "this generalization to one after another of the fundamental human interests, to show how it affected the boundaries of political divisions, the scope and nature of collective organizations, working loyalties and educational necessities." Wells attributed the introduction of the idea to Grant Allen "in an unpretending essay on the distances between country towns."

Among geographers the theme has been consistently followed in Britain by R. E. Dickinson and latterly by A. E. Smailes. Thus, the former studied the market towns of East Anglia, noting their distribution and range of functions in the 16th century and the subsequent competition between them and their thinning out in succeeding centuries. This and similar work in Germany and America have recently been reviewed by Dickinson in a valuable summary. There has, indeed, grown up a live "urban geography," an important and indeed indispensable sub-division of social geography, which bids fair to provide us with some of the generic concepts which are so deficient in the earlier "human geography."

Most, but not all, towns may be regarded as in some sense the functional centres of regions. If they fail to show this quality, they lack the most vital characteristic of a live town and must be degraded to the status of mere "urban tracts" as are some mining settlements and resorts. These latter occupy part of a region, but show no fully developed town-country relation. Chief interest centres upon the large cities which have an evidently "metropolitan" character and which are cardinal to large regions. The "field" or sphere of influence of such a city can clearly be investigated under a number of different heads. We may examine the "community range" of its working population, the local sources of its food supply, the radius of its diverse distributional functions, the regions served by its banks, insurance companies, hospitals and educational institutions. The several perimeters thus defined will not, of course, necessarily coincide, but they may do so sufficiently to enable us to define the urban sphere in a useful way.

Thus, Dickinson in a study of the twin cities of Leeds and Bradford showed that the former is not the functioning capital for the whole of the county of Yorkshire. Parts of south Yorkshire look to Sheffield as "town," whilst north of Northallerton the major functioning regional capital is Newcastle, although sub-capital functions are obviously exercised by the Tees-side group of towns.

The same line of thought has been carried further by C. B. Fawcett and latterly by other writers, in seeking to divide England into provinces each with its provincial capital. In the earlier version of his scheme, Fawcett recognized London, Southampton, Norwich, Oxford, Bristol, Plymouth, Birmingham, Nottingham, Manchester, Leeds, Sheffield and Newcastle as regional capitals. He defined his provincial boundaries exactly, choosing them so as to interfere as little as possible with ordinary movements and activities of people, to leave all parts of a province easily accessible to the capital, to render no province so populous as to dominate the rest, and to leave the smallest with a population sufficiently large for self-government. Attention was paid to the facts of local patriotism and tradition and, in general, the boundaries were drawn on or near watersheds. The largest of his areas, the London province, had a population in 1911 of over 10 million people, the smallest, "Peakdon" or the Sheffield province, did not greatly exceed one million. He has since slightly modified his scheme, absorbing the Peakdon province into the Yorkshire province, and modifying the boundary between those of Birmingham and Nottingham (Fig. 10).

There is something very attractive about the concept of a region related to its capital city, its integrating focus. Here surely, it may seem, we are thinking in terms of an *organized* whole. Yet although this idea is patently a valuable one, it can in no sense be taken as the heaven-sent solution of the regional problem. It carries with it a risk of wholly erroneous argument from analogy. We offend no usage, it is true, by calling such a region "organized" but to regard it as in any sense as an organism, subject to biological laws, is as fantastic as to regard each city as a "centre of crystallization." The phenomena in fact obey neither biological nor physico-chemical laws. We may borrow a certain amount of "pretty" metaphorical language from seeming analogies, but the phenomena are in fact wholly "social."

It is clear that, by implication, a large extension of the argument is involved when we pass from urban spheres,

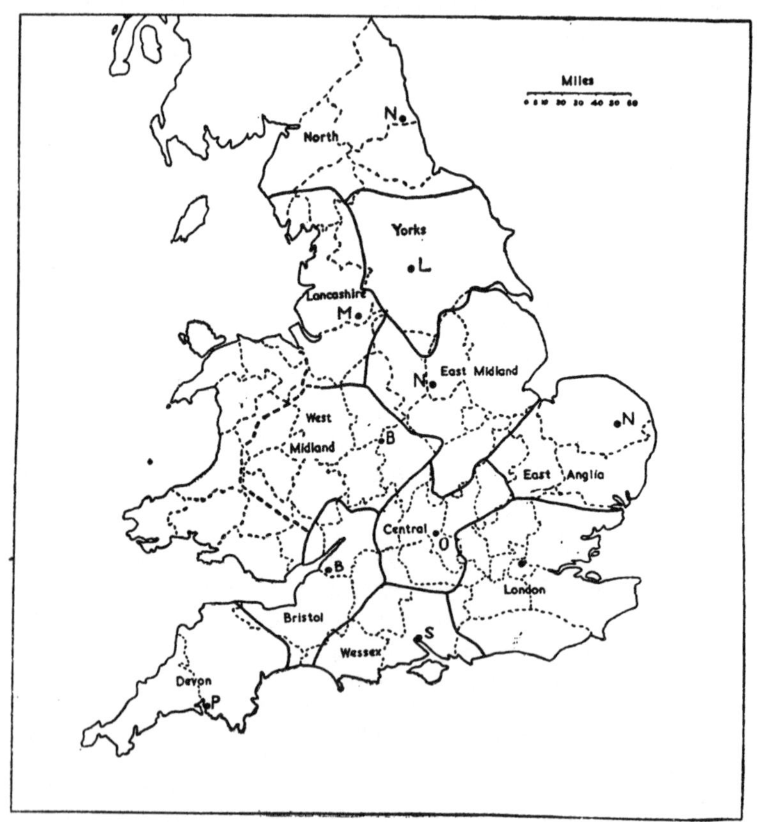

Fig. 10. The "Provinces" of England (after C. B. Fawcett).

quantitatively and statistically determined, to larger regions or provinces of which the same towns can be regarded as *ad hoc* capitals. In England, it is true, Fawcett's capitals, taken as a group, appear plausible and convincing. Yet the strength of their liaison with their regions varies widely. London, Birmingham, Manchester and Leeds are fully and truly metropolitan, Oxford and Southampton far less so. Even Norwich, although in some ways an excellent example of a regional capital, does not focus East Anglia in truly metropolitan fashion.

It is clear from Fawcett's announced criteria that the determination of his regions is on a basis which, if perfectly logical, is definitely special. Oxford and Southampton are forced by the logic of the scheme to be assigned a metropolitan rôle, because to include their province with large surrounding regions would make these latter over-large in area, population or both. It is clear indeed that the process of urban integration, if we may so term it, is new and nowhere necessarily complete. If we work outwards from urban centres defining by any suitable group of criteria the radius of their spheres, we shall not get wholly definitive boundaries, nor shall we be able to share the whole of the available area between them, except as a matter of rough convenience. The latter difficulty is not, indeed, acute in Britain and western Europe. Its reality is much more obvious in such an area as the United States.

The following cities have been claimed as metropolitan centres: Portland, San Francisco, Los Angeles, Salt Lake City, Denver, Minneapolis–St. Paul, Chicago, St. Louis, Dallas, Detroit, Cleveland, Cincinnatti and Atlanta, Boston, New York, Philadelphia and Baltimore. It is at once clear that these serve as centres for areas differing almost extravagantly in geographical character and, above all, in density of population. Here are several "Oxfords" and "Southamptons," the heads not of relatively small "relict" regions but of vast thinly peopled territories.

In South America the case is even clearer. The population map shows distinct clusters, for the most part widely separated, and each with its urban nucleus. As Preston James indicated, the pattern is an elementary one in an early stage of development. The clusters have not grown into contact, nor do the functions of their urban centres overlap. So empty is the heart of this continent that, although a study based simply on the clusters might adequately cover the "peoples," it would leave unaccounted most of the "lands" of South America. Even more would a study of Africa based entirely on its major cities prove geographically incomplete.

From this brief discussion it is clear that we cannot organize the whole world into regions marked by truly metropolitan foci. Only in Europe west of the Volga and North America east of the 100th meridian and to a smaller extent in Monsoon Asia is the urban network in any sense close enough for this purpose. This reflects the self-evident fact that the degree of development (in the metaphorical photographic sense) differs from region to region. Professor A. Stevens has noted an analogy between this idea and the plant ecologist's notion of "climax vegetation." An area cleared of plant life shows a succession of stages of colonization leading to the climax community—the most complex and fully developed which the area can support. We must not press this hological analogy too far, but it is none the less valid and illuminating.

It might therefore appear that the filling out of the "city regional" pattern was merely a question of time. In fact this is doubtless true but recent current tendencies in population growth and movement by no means wholly support this idea. The enormous growth of population in the last two hundred years has tended to aggregate itself ever more narrowly in large towns, town groups or urban regions, leaving the formerly empty quarters of the earth still relatively empty. It is a veritable "swarming" of our species which the population maps reveal, an evident response to industrialization and to the smaller amount of labour necessary to produce the requisite supply of food and agricultural raw materials. Notable exceptions to this generalization might doubtless be found, as formerly in Russia, but inter-census demographic trends in the Soviet Union between 1926 and 1939 showed a marked rise in the proportion of town-dwellers—from 24 per cent in 1926 to 33 per cent in 1939—the result not only of increasing population but also of its redistribution attendant upon planned industrialization. The fact of "swarming" is broadly true and certainly significant. It is indeed within the legitimate range of imagination to conceive of new tendencies and a reversal of present trends. A geographer writing an *Anticipations* in 1950 would inevitably speculate on the effect of new resources of power, and could not fail to take into account the current ideas upon the deliberate control of the growth of towns and the calculated "plantation" of new towns.

It is then obvious enough that in its present state of development we cannot divide the whole of the land surfaces of the world into "city regions" upon any consistent basis. For this reason alone we need constantly to employ the method of analysis which distinguishes "regions of uniform rural or natural pattern." Not only do these avail us when the other method is inapplicable but they remain significant *within* the city region. This is most clearly seen if we consider such a region as a "planning region." It is no doubt true that a great city, by reason of its physical extent and radial route system which links it to its region, obliterates or transforms older geographical patterns. Not only, however, do these older patterns exercise some influence on the form and functions of the new urban settlement, but in so far as they once reflected real physical differences, they still point to factors which will be relevant in planning at least the land use of the region.

It is, in other words, quite clear that whether for purposes of geographical description or of practical policy, a division of such a country as England simply and solely into "city regions" would be naïve and lop-sided. It is against this tendency so to regard matters that the geographer has constantly to warn the other social scientists and those of his colleagues heady with the new wine of city regions.

Let us imagine a conducted tour of a party of foreign geographers or observant laymen, unacquainted with England. As our visitors traversed, say, the highly characteristic Cotswold countryside, they might pass through parts of the Bristol, Birmingham and Oxford "provinces." This would be far from obvious however: the metropolitan foci would be below the local horizon and the unities and harmonies of the rural landscape would be far more impressive. The life of the Cotswold region is indeed affected by the various distant urban foci, but it is also affected by what might be called the "indigenous characteristics of the terrain." Those geographers indeed who insist that "landscape" in its fullest sense is the core of geographical study would seem to have no choice but to accord high place to regions of the Cotswold type. This is evidently not to say that city regions are either less important or unimportant. The two regional conceptions are not antagonistic but complementary: we cannot fully appreciate regions of one sort without considering more of the other.

It is by no means merely fanciful and in many ways practically helpful to regard our two types of region as the continuing expression of the two great social and economic revolutions of mankind to which prehistory bears witness. The essential elements of the "humanized" rural landscape, despite all later transformations and elaborations, date from the neolithic revolution—the advent of farming. The town and all that it stands for in the specialization of labour, the differentiation of classes, transportation, trade, etc., began essentially with the urban revolution of the Bronze Age.

In Europe at least urban life and sedentary rural life have co-existed in varying degrees of isolation or mutual relation since the beginning of the third millennium B.C., spreading from their region of origin in the Near East. In some areas especially, and in considerable degree throughout the greater part of the world, we are brought forcibly to recognize the geographical imprint of the third great transformation—effected by the so-called Industrial Revolution which brought about acceleration and perhaps hypertrophy in the urban tendency, changing it in degree if not in kind.

To see these phases in broad historical succession is better to understand the complexity of the geographic patterns. The urban element always appears at first sight something alien and cohesive grafted upon the rural landscape: it has its own distinctive patterns and processes. But this is really to overstate the case: we cannot artificially divorce country from town in geographical study. We are witnesses rather of a complex and gradual process of "symbiosis" whereby the developing life of the city and the city-region transforms the character and subsumes the categories of rural life. Nevertheless there is virtue in the characteristic French fashion of regional description used by Vidal de la Blache and his school. The urban life of an area is specially treated after its general landscapes have been surveyed: the order of geographical presentation follows the order of historical development.

The reader will have become aware that the study of regional geography, like present-day landscape in its physical and human aspects, has reached only a stage in its development. The purpose of regional geography is simply the better understanding of a complex whole by the study of its constituent parts. It will have become clear that there is no single route open to the regional geographer. Since geography is concerned with both warp and woof, with both physical setting and human impress, the search for regional differences of many kinds can, indeed must, be attempted. Analogy may be drawn from the study of society.

The sociologist is concerned with numerous societies of different types and scale, such as family, school, profession and nation, of which the same individual may form part, and all of which are comprised in the Great Society of mankind. So also, for the geographer, the same area of land may form part of a whole succession of regions of different scale and character—physical regions, industrial regions, urban regions, administrative regions and others—of which the earth's surface as a whole is comprised. Coincidence between the boundaries of such regions would only exceptionally occur, and the full understanding of an area would involve the discovery of its relationship to regions of the many kinds. To the regional geographer, therefore, the landscape presents a jig-saw puzzle with several sets of pieces; nor, as we noted above in discussing urban regions, do these pieces always neatly interlock.

The critic would be wrong who averred that regions exist solely in the minds of geographers for, to a large extent, they are visible and demonstrable realities. We have focused attention here on regions of two types, the detection of which and their delimitation on maps, present clearly a heavy though rewarding task to the geographer. These two types claim the attention of geography as a whole as distinct from its specialist branches.

Urban regions spring from human settlement and social relationships in area; although in some measure visible landscape forms, they are essentially dynamic in character, in response notably to changing social purpose and to the pace and facilities of transport. They permit us to understand one main aspect of social geography—the areal relations of country and town, but leave aside other social relationships in area, for example, those between rural settlements (e.g., nucleated villages and scattered farms).

While urban regions depend for their existence on social functions and are pre-eminently man-made units, regions of the other type, based on the association of "sites," themselves environmental cells, are solely physical divisions, for they are determined by reference to purely physical considerations. In theory, the whole land surface of the earth could be shown to be composed of regions of this kind—to be in fact a gigantic mosaic of irregularly patterned shapes. These regions are wholly visible and demonstrable in the field, although an element of subjective judgment enters into their delimitation on the map. Because it is determined in terms of rock outcrop, soil and climate, the physical region is, in contrast to the urban region, a virtually static entity: the physical changes to which it is subject are not often measurable within narrow human periods. For these reasons, and because they are significant also in relation to urban regions, the physical region may perhaps claim primacy in the hierarchy of geographical regions. Social groups, it is true, will develop these inert physical environments, according to their ability and will, and recast them as constituents of human regions of many kinds. Yet if, as we have argued in this book, the central objective of geography is the study of country rather than of man, the physical region is the protagonist in the geographical arena.

CHAPTER IX

CONCLUSION

Geography begins only when geographers begin writing it.

WE began by quoting the formally correct but rather misleading dictionary definition of geography. Everything in the foregoing chapters is, in effect, an expansion and commentary upon that definition and we have left much unsaid. Many other aspects of a great and growing subject present themselves for attention and still others will be discovered as the scope of geographical thought is extended and consolidated. All of them stem from the central root which, expressed in the simplest terms, is that the earth is the home of man and that man, whatever else he may be, is at least one with the earth and part of nature. The disputations about the scope and status of geography start at this point. To say that the earth is the home of man and man part of nature may, on the one hand, be dismissed as a palpable truism or, on the other, regarded, as the geographer regards it, as a profound truth worthy of detailed study and careful reflection.

In seeking here to draw together a few of the lines of thought we have sought to follow, we note in the first place that geography as a subject involves not special material and a rigidly bounded field but a point of view. Lest this seem an over-modest and insufficient claim, let us recall that it is not a point of view lightly adopted for mere purposes of argument but one only to be attained by an arduous discipline. To the study of many human problems the lawyer brings a distinctive and indisputable point of view, based on a recognized expertise. No one is likely to underrate the value and validity of legal training because it is applied to problems with elements common to many other fields. So it is, and increasingly must be, with geography.

The geographer is perfecting a tool, one of many, which avails in the attack on a great range of political, social and economic problems. Some may prefer, and legitimately so, to emphasize this aspect and see the justification of geography in its contribution to statecraft in the widest sense, and in man's ceaseless efforts consciously to control his terrestrial environment. There has indeed arisen in recent years a veritable applied geography in the field of town and country planning. For the first time the press has published advertisements for geographers to assist in what are essentially problems of land use. Many would claim with good show of reason that the major contribution of British geographers in recent years has been the completion of the Land Utilization Survey and the cognate enquiries since carried out in the various technical branches of the Civil Service. This work will in no sense stand or fall by the verdict of history or of political philosophy on the propriety and success of a "planned economy." Applied geography moves towards no predetermined conclusions and offers no premature verdict on the fiercely debated question of "planning" in the large.

The attitude of the geographer in this field is scientific in the best sense: it involves a cool appraisement and measurement of what is, and the vivid imagining of what might be, without any initial bias towards either detailed means or ultimate ends. It is true that such absence of bias, once universally sought as a virtue in the world of learning, is now bitterly assailed by an influential group of British scientists. There is some force in their castigation of the more ludicrous manifestations of academic isolation, but great danger in their facile assumption that the end is known and the means of reaching it obvious.

But by the very nature of his subject, the geographer is protected from the perils of isolation and specialization. Not for him the characteristic "one-eyed" approach, which has, it is true, its limited value if properly controlled, but carries with it the danger of producing a permanent squint. You may charge him ever and again with too wide a view or too ready a generalization, and against these propensities he must be continually on his guard. But that the subject requires wide views and should attempt generalizations is not in doubt: what matters is that they should be based on patient detailed work carried out with the thoroughness of true scholarship. The recurrent temptation to which the geographer is exposed in his more sanguine moments is to arrogate to himself a sort of supervisory rôle over other specialists, as if to say that if they will do the work he will draw the conclusions.

This prompts the very just retort that he is attributing to himself high if not superhuman qualities of mind that he clearly does not possess. His real claim is, more modestly, that he is attempting to "see things together" and that such seeing is an art not to be acquired without cultivation and training. The separate strands of his pattern pass beneath the plane of the immediate and the obvious, but the pattern none the less exists, the whole is greater than the sum of the parts and an effort must be made to view it in its entirety. The present need for integration in the divergent and multiplying fields of human knowledge is urgently acclaimed. Geography offers such an integration over part of the field and the character of its spirit and the manner of its service must be judged in the light of this fact. The claim for geography indeed depends fundamentally not upon its application but upon the necessity of careful study of evidently significant phenomena in the field of earth and man.

It is an illusion to suppose that a just geographical view emerges spontaneously from the separate labours of geologists, historians and others. The "binocular" principle must be deliberately applied to the intellectual synthesis, with the proviso that it is not two but many aspects which must be brought into related review. Only those who, either through ignorance or prejudice, perversely deny these relations can condemn the only method by which they can be perceived.

All this may be fit meat for the professional geographer and his critics and yet seem rather inadequate fare for the interested layman or the student beginning his work. For them the best approach is the practical one. It is a curious and significant fact that geography has so far bred few amateur observers. In many tracts of the English countryside the life of the pond has been diligently surveyed, the fossils collected from quarries and lane banks; quite independently village histories have been written and parish churches studied. That there is any regional relationship between these and many other disparate strands of knowledge is either ignored or taken for granted as a fact of little interest or significance.

Great studies like the *Victoria County History* provide material for geography but the synthesis is not made unless the geographer attempts it. Yet no great feats of learning and scholarship, impossible of attainment, are involved. The observant eye and the reflective mind, providing it is map-conscious, can achieve real additions to knowledge. In further fields the naïve observations of the untutored traveller are grist for the geographer's mill. We judge these adversely by reason of the fact that too often relations are not sought and therefore not seen. Again, many a district commissioner in colonial territories has under his hand, or even in his mind, a vast knowledge which can be made to live and cohere by the art of the geographer. In its simpler modes the art perhaps seems over-simple and the specialist believes, correctly perhaps, that he can achieve it himself if he is interested enough to take the trouble. Yet he rarely makes the attempt.

The more elaborate art may perhaps produce unworthy irritation in that the new method seems to bring order out of chaos. Neither geologist nor historian in fact produced the luminous account of Britain found in *Britain and the British Seas*: it remained for the geographer Mackinder to write it. No other than the geographical method applied by de Martonne could have produced so well-balanced and penetrating a survey of Central Europe, nor could either a sociologist or an economist have produced a substitute for the *Tableau de la Géographie de la France* of Vidal de la Blache. To such the student of the subject or its critics should turn before dismissing it as an *omnium gatherum*.

The further fundamental fact of which sight is too often lost is that geography begins at home. One may fairly suspect the pretensions of the geographer who cannot or does not interpret the country in which he lives. It is true that, in another sense, the world is his unit and, in pursuance of his own principle, the study of the whole assists the interpretation of the parts. But the geographical method is seen at its best where the data are fully available and the ground accessible to study.

We have admitted that the trained geographer working with his maps can often present a sounder picture of a distant land than those who know it intimately in some limited practical context. But this is no argument for ignorance of the ground. A taste for geography is no doubt often fostered by reading the narratives of travel in far-off lands and strange places. It is still more securely and fundamentally based upon the experience of wandering, map in hand, in one's native countryside.

In a mood of disillusionment, overcome by the complexities of the problems, the geographer on occasion is minded to assent to the attacks of the critics and to concede that theirs is the better part who select for study one aspect only of this world of men and things. The best cure for this mood is to go once more into the field and savour once again the unity of man and nature and the correlation between physical and social phenomena which confront him on every side. It is then that he realizes anew that the proportions and relations of things are as much facts as the things themselves and that, in the geographic field, unless he studies them, no one else is likely to do so. His subject, no less than others in the curriculum, subjects him to a discipline and yields him a philosophy.

NOTE ON READING

This short note is designed to indicate elementary books with which a beginner may fittingly start—these are marked with an asterisk—and to give further particulars of works referred to in the text.

CHAPTER I

THE NATURE AND DEVELOPMENT OF GEOGRAPHY

R. Hartshorne, "The Nature of Geography," *Annals of the Association of American Geographers*, Vol. XXIX, Nos. 3 and 4 (1939).

*H. J. Wood, *Exploration and Discovery* (1951).

CHAPTER II

THE PHILOSOPHY AND PURPOSE OF GEOGRAPHY

P. Vidal de la Blache, *Principles of Human Geography* (English trans.), 1926.

C. O. Sauer and J. B. Leighly, *Introduction to Geography* (Part 1, 1932).

V. C. Finch and G. T. Trewartha, *Elements of Geography*, New York, 3rd ed., 1949.

J. Brunhes, *Principles of Human Geography* (new English edition, Harrap, 1952).

C. Daryll Forde, *Habitat, Economy and Society*, 6th ed., 1948.

Vaughan Cornish, *The Great Capitals* (1923).

CHAPTER III

PHYSICAL GEOGRAPHY AND BIOGEOGRAPHY

P. Lake, *Physical Geography* (new revised edition), 1949.

S. W. Wooldridge and R. S. Morgan, *The Physical Basis of Geography*, 1937.

J. E. Marr, *The Scientific Study of Scenery*, 9th ed., 1943.

S. Petterssen, *Introduction to Meteorology*, 1941.

W. G. Kendrew, *Climatology*, 1950.

A. G. Tansley, *Britain's Green Mantle*, 1949.

M. I. Newbigin, *Plant and Animal Geography*, 1936.

CHAPTER IV

GEOGRAPHY AND MAPS

*F. Debenham, *Map Making*, 2nd ed., 1940.

*Col. Sir Ch. Close, *The Map of England*, 1932.

A. R. Hinks, *Maps and Survey*, 5th ed., 1944.

CHAPTER V

HISTORICAL GEOGRAPHY

*E. G. Bowen, *Wales : A History and Geography*, 1st ed., 1940.

*W. G. East, *The Geography Behind History*, 1st ed., 1938.

L. Febvre, *A Geographical Introduction to History* (English trans.), 1925.

Graham Clark, *Archaeology and Society*, 2nd ed., revised, 1947

V. G. Childe, *Scotland Before the Scots*, 2nd ed., 1946.

Sir C. Fox, *The Personality of Britain*, 4th ed., 1947.

H. C. Darby (editor), *The Historical Geography of England Before 1800*, 2nd ed., 1948.

W. G. East, *An Historical Geography of Europe*, 4th ed., 1950.

Ordnance Survey Maps of Roman Britain (2nd ed., 1932), and of 17th–Century England and Wales (1930).

CHAPTER VI

ECONOMIC GEOGRAPHY

L. Robbins, *The Nature and Significance of Economic Science*, 2nd ed., revised, 1935.

*M. I. Newbigin, *Commercial Geography* (Home University Series), 2nd ed., 1928.

*C. A. Fisher, "Economic Geography in a Changing World," The Institute of British Geographers, *Transactions and Papers*, 1948 (Philip and Son).

R. O. Buchanan, *The Pastoral Industries of New Zealand*, Institute of British Geographers, 1935.

Sir Josiah Stamp, "Geography and Economic Theory," *Geography*, XXII, 1937.

*W. R. Mead, "Agriculture in Finland," *Economic Geography*, April-May, 1939.

E. W. Gilbert and R. W. Steel, "Social Geography and its Place in Colonial Studies," *Geographical Journal*, CVI, 1945.

CHAPTER VII

POLITICAL GEOGRAPHY

*A. E. Moodie, *The Geography Behind Politics* (Hutchinson's University Library), 1949.

Sir H. J. Mackinder, *Democratic Ideals and Reality*, 1919 (Penguin edition, 1944).

S. W. Boggs, *International Boundaries*, New York, 1940.

*H. W. Weigert and V. Stefansson, *Compass of the World*, 1945

*H. W. Weigert, V. Stefansson and R. E. Harrison, *Second Compass of the World*, 1949.

Y. M. Goblet, *The Twilight of Treaties* (English trans.), 1936.

W. G. East, "The Nature of Political Geography," *Politica*, 937.

W. G. East, "The Political Division of Europe," Birkbeck College, 1948.

W. G. East and O. H. K. Spate (editors), *The Changing Map Asia*, 1950.

R. Hartshorne, "The Functional Approach in Political Geography," *Annals of the Association of American Geographers*, XL, 2, June, 1950.

CHAPTER VIII

REGIONAL GEOGRAPHY AND THE THEORY OF REGIONS

C. B. Fawcett, *The Provinces of England*, 1919.

R. E. Dickinson, *City, Region and Regionalism*, 1947.

A. E. Smailes, "The Urban Hierarchy in England and Wales, *Geography*, XXIX, 1944.

A. E. Smailes, "The Urban Mesh of England and Wales, Institute of British Geographers, *Transactions and Papers*, 1946

Preston E. James, *Latin America*, 2nd ed., 1947.

H. R. Mill (editor), *International Geography*, 4th ed., 1907.

INDEX